SpringerBriefs in Applied Sciences and Technology

T0235646

Series editor

Andreas Öchsner, Southport Queensland, Australia

For further volumes:
http://www.springer.com/series/8884

Olavi Uusitalo

Float Glass Innovation
in the Flat Glass Industry

 Springer

Olavi Uusitalo
Tampere University of Technology
Hollola
Finland

ISSN 2191-530X   ISSN 2191-5318   (electronic)
ISBN 978-3-319-06828-2   ISBN 978-3-319-06829-9   (eBook)
DOI 10.1007/978-3-319-06829-9
Springer Cham Heidelberg New York Dordrecht London

Library of Congress Control Number: 2014938211

Printed on acid-free paper

Springer is part of Springer Science+Business Media (www.springer.com)

Foreword

Industries are man-made. They develop and change. They create opportunities and influence our lives. In this book, Prof. Olavi Uusitalo in a very insightful way tracks, describes and analyses the impacts of technological change on the structures of the flat glass industry in the U.S. and American markets with particular focus on the emergence and diffusion of float glass technology in two independent subindustries, sheet glass and plate glass in the markets mentioned above.

In this work, Uusitalo in a novel and an insightful way employ multiple data, including among others data from business magazines and interviews. In this way, he succeeds in getting more nuanced and detailed insights into the industry studied than what is usually the case. Also—his detailed analyses are insightful.

This work is no doubt of value to others and is highly recommended to students and others who are interested in industry development and change, industry analyses, as well as technology change and impact.

Bergen, March 2014   Kjell Grønhaug

Preface

Since the 1930s research has concentrated on the impacts of technological change on industries. They have been studied in several disciplines such as industrial economics, organizational research, and population ecology. Today technological change is more rapid and thus research concerning its effects on industry structures is extremely important. This study is also a part of this tradition and is based on the literature stream of the cyclical model of technological change or the dominant design model. This stream is widely cited by scholars from several disciplines. The main objective of this study is to evaluate whether the number of digits of the Standard Industrial Classification (SIC) code or equivalent in the definition of industry has any meaning. In the base study of cyclical model of technological change, the industry was defined with four-digit SIC code. I argue that the definition of industry with four-digit SIC code disaggregates in the case of the flat glass industry, one of the studied industries, two important subindustries and one crucial flat glass processing industry, the safety glass industry, from the analysis. Moreover, the international aspect so important to both the subindustries, the plate glass and window glass industries, of the flat glass industry are treated too narrowly or even neglected, as well. This study uses five-digit SIC codes which allows deep enough analysis of the relevant sub and value-added industries and the industry-specific international aspects. This book reveals the difference within these two approaches.

This study explores the impacts of technological change on the structures of the flat glass industry in the U.S. and European markets. Empirically, this research focuses on the emergence and diffusion of float glass technology in two independent subindustries, sheet glass and plate glass of the flat glass industry in the two above-mentioned markets. A radical technological change, float glass, introduced in the plate glass industry in 1959 was quickly adopted by the plate glass industry, where it eliminated the labor- and cost-intensive grinding and polishing production phases. Sheet glass was a cheaper product and its manufacturing process did not require as huge capital investments as were needed in the plate glass industry. However, float glass gradually entered the sheet glass markets as well and finally the float glass process replaced the existing drawing methods. The research period in this longitudinal case study is 60 years (1920–1980).

The empirical findings of this study and those of the base study of the cyclical model of technological change are not the same. Some of the differences seem to be based on the definition of the industry that is the number of digits in SIC codes. The four-digit SIC code reveals only the flat glass industry while the five-digit SIC codes reveals also the two subindustries, the plate glass and sheet glass industries plus the flat glass processing industry, the safety glass industry so pertinent for plate glass. In the base study the four-digit SIC codes may have been the reason for the problematic and inconsistent interpretation of the nature of the gradual convergence of two subindustries. After redefinition of three major concepts of the dominant design model, the modified model is re-examined and re-tested in the flat glass industry in the period of 1950–1980.

Good understanding of the industry is utmost important in historical industry studies. The present report contributes to industry studies by using extensive qualitative analysis as the methodological approach. Also the definition of industry by five- and even six-digit SIC codes made the analysis more thorough. The re-examination suggests two consequences of dominant design, a gradual convergence of the two subindustries and market concentration after a competence-destroying technological change. This study has increased our understanding of the impacts of a technological change on the industry structure.

Acknowledgments

I would like to express my gratitude to people who have contributed to my research in the flat glass industry since its initial stage. First of all, I would like to express my special thanks to Prof. (Emeritus) Kjell Grønhaug from Norwegian School of Economics. The discussions, encouragement, and support from him during the 20 years co-operation have been of great help. I would like to thank Prof. Kristian Möller for his valuable advice, comments, and encouragement in the early phase of my research. Prof. Erkki Pihkala has also guided me in my research.

I would also like to express my gratitude to the people helping me to get in the flat glass industry in the early 1990s. First of all, I would like to thank Sir Antony Pilkington, chairman of Pilkington that time, for his positive attitude toward my research. He also provided with valuable information and comments. I also would like to thank the top management of Lahden Lasitehdas and Lamino Group for their help during my research. I would like especially to express my gratitude to Jonas Borup of Lahden Lasitehdas, MD that time and Erkki Artama of Lamino, MD that time, for their support at the beginning of the empirical part of this research. Furthermore, both of them provided me with excellent sources of information. Without the excellent histories of Pilkington and the flat glass industry written by Prof. (Emeritus) T. C. Barker of London School of Economics and Political Science, this research would hardly have been possible. I would also like to thank Prof. Barker for his comments, advice, and encouragement in the mid-1990s.

I would like to thank Mr. Toni Mikkola, research at Tampere University of Technology, Industrial Management for the discussions and comment on the later works relating to the float glass innovation. I would like to thank Anthony Doyle, Senior Editor Engineering, at Springer providing for his interest on my research on the flat glass industry.

I would like also to express my thanks to my wife Kaija and my children Laura, Saara, and Kari for the support and understanding I have received from them during the a very long time. I would like to dedicate this piece of research to wife, Kaija.

Olavi Uusitalo

Contents

Abbreviations

AFG	AFG Industries (the successor of ASG Ind. in 1978)
AGR	American Glass Review
Asahi	Asahi Glass Co.
ASG	American St. Gobain
ASG Ind.	ASG Industries (the successor of ASG in 1970)
AWG	American Window Glass
Boussois	Glaces de Boussois
BSN	Boussois-Souchon-Neuvesel
CCCN	Customs Co-operation Council Nomenclature
C-E	Combustion-Engineering Inc.
Delog	Deutsche Libbey-Owens-Gesellschaft für Maschinelle Glasherstellung
Detag	Deutsche Tafelglas AG
FDI	Foreign Direct Investment
Ford	Ford Motor Co.
Guardian	Guardian Industries
HMS	Ateliers Heuze Malevez et Simon
LOF	Libbey-Owens-Ford Glass or LOF Industries
MD	Managing Director
MGU	Multi Glass Unit
MNC	Multinational Corporation
PB	Pilkington Brothers Ltd.
Pilkington	Pilkington Brothers Ltd./Pilkington plc
PPG	Pittsburgh Plate Glass or PPG Industries
R&D	Research and Development
S.I.V.	Societa Italiana Vetro
SIC	Standard Industrial Classification
St. Gobain	Compagnie de St. Gobain
TC	Tariff Commission
TDS	Thick Drawn Sheet Glass
TGI	The Glass Industry Trade Journal
Vegla	Vereinigte Glaswerke GmbH

Chapter 1
Introduction

Abstract This chapter discusses the literature on technological change and industrial change culminating in the introduction of cyclical model of technological change. It also brings up some weaknesses in the operationalisation of the model. The chapter also illustrates the float glass innovation, one of the most brilliant innovation in the last century and how it converged two independent sub industries to one industry. At the end of the chapter the construction of the analysis is shown.

Keywords Dominant design · Flat glass industry · Float glass · Technological change

Since the pioneering work of Schumpeter (1934, 1942), research has concentrated on the effects of technological change on industries. The effects of technological change are studied in several disciplines. According to Suarez (2004) management scholars have used the following labels: "dominant designs" (Utterback and Abernathy 1975; Anderson and Tushman 1990; Utterback and Suarez 1993), "technological trajectories" (Dosi 1982; Sahal 1981) and, more recently, "platforms" (Meyer and Lehnerd 1997; Cusumano and Gawer 2002). Suarez (2004) continues that economists have studied the same phenomena under the labels of "standards" in network industries (Katz and Shapiro 1992; David and Greenstein 1990); and "technology diffusion" (Reinganum 1981), "standards wars" and network effects (Arthur 1998)—although some of their claims have since been disputed (Liebowitz 2002). Since the invention of the first semiconductor in the 1950s, technological change has been extremely rapid. Thus, research in this area is dominant design and its diffusion in the industry.

O. Uusitalo, *Float Glass Innovation in the Flat Glass Industry*,
SpringerBriefs in Applied Sciences and Technology,
DOI: 10.1007/978-3-319-06829-9_1, © The Author(s) 2014

1.1 Literature on Technological Change and Industrial Change

Many scholars from different disciplines have contributed to the literature which combines technological change and industrial change. I have divided this type of literature into three groups: industrial economists, organizational researchers, and population ecologists and evolutionists. The first category includes scholars such as Schumpeter (1934, 1942), Freeman (1982), Utterback and Abernathy (1975), Abernathy and Utterback (1978), Abernathy (1978), Sahal (1981), Nelson and Winter (1982), Dosi (1982), Porter (1983), Pavitt (1984), Abernathy and Clark (1985), Clark (1985), Foster (1986), Dosi et al. (1988), Teece (1986, 1988), and Utterback (1994). The second group includes the work of Tushman and Romanelli (1985), Tushman et al. (1986), Tushman and Anderson (1986, 1987), Anderson (1988), Romanelli (1989), Anderson and Tushman (1990, 1991), Tushman and Rosenkopf (1992), Rosenkopf and Tushman (1994), McGrath et al. (1992) and Tuhman and Murmann (1998). The most influential authors in the third group are Hanna and Freeman (1977, 1989) and Britain and Freeman (1980).

In the 1980s and 1990s a growing number of scholars used biological analogies throughout the social sciences to explain the dynamics that govern technical and institutional changes as a social evolutionary process of variation, selection, and retention (see Fig. 1.1 Van de Ven et al. 1994 Fig. 20.1). Variation means technological or institutional change. Selection or emergence of a dominant design occurs principally through competition among alternative novel forms (technology in the present study). Retention involves the forces that try to maintain certain technical and institutional forms that were selected in the past. Van de Ven et al. (1994:426) classify the authors presented above and indicate the popularity of Anderson and his associates' model as follows:

"Organizational scholars view the temporal relations among variation, selection and retention either as a continuous and gradual process (McKelvey 1982; Nelson and Winter 1982; Hannan and Freeman 1989) or as a punctuated equilibrium (Utterback and Abernathy 1975; Tushman and Romanelli 1985; Lumsden and Singh 1990). The latter and currently more popular view, is exemplified by Anderson and Tushman's (1990) cyclical model of technological change illustrated in Fig. 20.1." (see Fig. 1.1)

Utterback and Abernathy (1975) for instance, contrary to Anderson (1988), assume that product innovations must always precede process innovations. Sahal's (1981) work mainly differs from Anderson's study in the way breakthrough innovations arise, how long-wave economic cycles are linked to secular trends in national technological development, and how technologies diffuse in space and time (Anderson 1988:29). In Sahal's model the role of the technological guidepost or standard in generating volume production is played down (Anderson 1988:31).

This classification by no means gives a clear-cut picture of the research traditions of technological and industrial change. The borders between these three groups are vague. However, it helped the present author to organize and

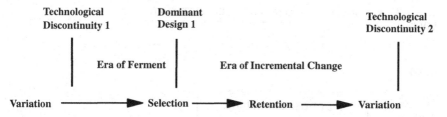

Fig. 1.1 Illustration of the Technology Cycle in the Punctuated Equilibrium Model (Reproduced from Van de Ven et al. 1994; adapted from Tushman and Anderson 1990)

conceptualize the research tradition. It seems, for instance, that the theoretical cornerstone of the present study, Anderson (1988), coming from the second group, organizational studies, also touches several topics from the other two groups, industrial economists and population ecologists and evolutionists. Romanelli (1989) seems to do likewise.

The present study belongs to the first group, industrial economics, although it borrows its theoretical base from the second group, organizational researchers. The topics of the last group, population ecologists and evolutionists, seem to be more or less beyond the concern of the present research.

The cornerstone of the present study, the cyclical model of technological change proposed by Anderson and Tushman (Tushman and Anderson 1986, 1987; Anderson 1988; Anderson and Tushman 1990, 1991), belongs to the organizational studies. Tushman and Anderson have borrowed the concept of dominant design from Utterback and Abernathy (Utterback and Suárez 1993:7). The term 'dominant design' was introduced by Abernathy and Utterback in 1975 (Suárez and Utterback 1995:416). According to Utterback and Suárez (1993), Anderson and Tushman (1990) have gathered valuable data on the minicomputer, the glass and the cement industries and have tested the model.

> "Their work not only provides more data to support the Abernathy-Utterback model, but also enhances the latter in several respects. They provide additional insights on the emergence of dominant designs, by looking at the problem as a political process (the authors have yet to explore this interesting idea fully). They also make an insightful distinction between competence-enhancing and competence-destroying discontinuities, an idea based on the taxonomy proposed by Abernathy and Clark (1985)". (Utterback and Suárez 1993:7).

Van de Ven et al. (1994) give merit to the Anderson and Tushman (1990) model by saying that the punctuated equilibrium model (see Fig. 1.1) provides a useful template for examining overall stages of coevolution, particularly across successive generations of technological growth.

Anderson and Tushman (1990, 1991) are extremely well cited. The former article has been cited 695 /2460 times (from Web of Science /Google Scholar) while the latter one has been cited 46 /215 times (from Web of Science/Google Scholar). Web of Science were read October 15, 2013 and Google Scholar March 21, 2014.

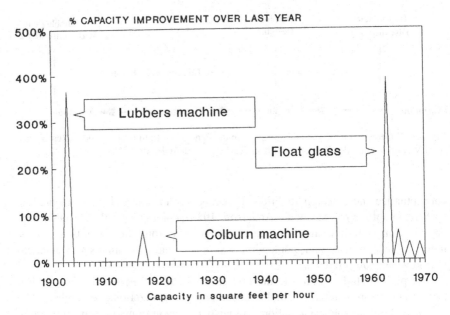

Fig. 1.2 Three great discontinuities mark the development of machinety for manufacturing window glass in the United States (Reproduced from Anderson and Tushman 1991)

1.2 Objectives of the Study

Anderson (1988) studied five different industries as seen from the following quote: "The study is based on an intensive historical study of three industries, one of which in turn (glass manufacture) contains three branches whose technology and markets are quite different." (Anderson 1988:8). However, although the window glass and plate glass industries were said to be different (see also Fig. 1.4) it seems that Anderson (1988) has studied flat glass industry (including both the plate glass and window glass industries) as a single industry.

The similar figures about technology progress were under the window glass (Fig. 1.2) in Anderson and Tushman (1991) and under the flat glass (Fig. 1.3) in Anderson and Tushman (1990). Figure C-1 in Anderson (1988:375) was under window glass. In both figures there are three technological discontinuities. Both Anderson (1988:371) in Table C-1 and Anderson and Tushman (1990:610) in Table 1 listed four technological discontinuities in the flat glass industry. They both have added continuous casting from the plate glass industry in the table. The tables also indicate that the flat glass industry has been studied as a single industry. Anderson and Tushman (1990:606) confirms the single study of the flat glass industry, since they have used four digit SIC codes according to the following quote:

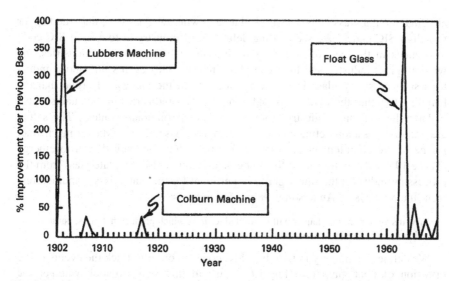

Fig. 1.3 Technological progress of machines in the U.S. flat-glass industry, in square feet per hour (Reproduced from Anderson and Tushman 1990)

"In this study, the technology of an industry was defined by its four-digit Standard Industrial Classification (SIC) codes to use standard industry boundaries" (Anderson and Tushman 1990:606).

In the flat glass industry the float glass innovation gradually converged the plate glass and window glass industries into a single industry as Anderson and Tushman (1990) noticed:

"Finally, decades of research culminated in the development of the float-glass process at Pilkington, a British glassmaker. Molten glass was passed across a bath of molten alloy; the production rate increased dramatically, since the ribbon was subject to less resistance in drawing. Additionally, the ribbon emerged perfectly flat, eliminating the need for grinding and causing the window and plate glass industries to converge into a single SIC code." (Anderson and Tushman 1990:607).

In fact Anderson (1988), Anderson and Tushman (1990) and Anderson and Tushman (1991) studied the flat glass industry (defined by four-digits SIC code = 3211) in which float glass converted the two sub industries into one. It is interesting that neither Anderson (1988) nor Anderson and Tushman (1990) regard float glass as an industry standard although it converged two industries.

The main objective of this study is partly derived from the inconsistencies in the number of technological changes in the flat glass industry reported in three publications by Anderson and Tushman (Anderson 1988; Anderson and Tushman 1990, 1991) and the fact that Anderson (1988) and Anderson and Tushman (1990) did not regard float glass as a dominant design although it converted the plate glass and sheet glass industries into a one single industry. As I saw Anderson (1988) and Anderson and Tushman (1990) define the industry by four-digit SIC codes although the flat glass' sub-industries plate and window glass were classified under five-digit SIC codes. We see that the definition of industry in four-digit SIC code aggregates in the case of the flat glass industry the relevant sub industries, the window glass industry and the plate glass industry plus even value added industries such as safety glass in the same category in the analysis. I argue that the handling together the plate glass and window glass industries in the analysis also isolates the flat glass industry from its crucial international context. The safety glass industry is a marketing channel to the plate glass industry. Marketing channel is one of the complementary assets so pertinent for dominant design to emerge (Teece 1986). Thus, four-digit SIC code disintegrated also the plate glass from the processing industry, the safety glass, so pertinent for the plate glass manufacturers from the analysis as Anderson (1988:335) confirms:

> "The discussion in this chapter will be limited to firms which produce flat glass, not those which produce products based on flat glass."

We define the industry in five-digit SIC codes in order to track the events in the diffusion of float glass (see Fig. 1.6). I try in this study to look whether the definition of the industry with four-digits (instead of five-digits) SIC code has an impact on the analysis of float glass. Thus,

the main objective of this study is to evaluate whether the definition of the industry by four-digit or five-digit SIC codes matters?

The research will be done by examining the definition of the industry within the concept of cyclical model of technological change proposed by Tushman and Anderson (Anderson 1988, Anderson and Tushman 1990, 1991) within the flat glass industry. As was mentioned Anderson (1988) and Anderson and Tushman (1990) used four-digit SIC codes thus leaving the plate glass and window glass industries out of the scope.

Empirically, this research seeks to increase our understanding first, of the impacts of a radical technological change on the industry structure generally and especially the impacts of float glass the flat glass industry structure. This study follows the encouragement given by Anderson (1988:194)

> "It is hoped that this study will show what can be done with the historical record, and encourage future scholars to confront falsifiable hypotheses with the data (and learn from the shortcomings of this research) when it makes sense to do so."

Based of our findings the definition of the industry with five-digit SIC codes makes a change. The solid analysis of the subindustries in the era of float glass reveals separate diffusion of float glass in both plate glass and sheet glass industries. With this definition the large U.S. flat glass manufacturers can be analyzed separately from the view point of both subindustries. Moreover I can conclude that float glass was dominant design in both subindustries.

1.3 Limitations

This study evaluates only whether the definition of industry either with four-digit or five-digit SIC codes makes any difference in the analysis. This study neither test hypotheses developed by Anderson (1988) nor the consequences of the application of the cyclical model of technological change or other results of Anderson (1988) and Anderson and Tushman (1990, 1991) by themselves of other scholars.

1.4 Research Approach

A research approach should be derived from the objectives of the study (Grønhaug 1994). This section describes briefly the research approach used in the present study. This section has three parts. The first part presents arguments for a single case study. The second part describes briefly the technological innovations occurred in the research period (1920–1980) in the flat glass industry. The float glass innovation, the technology to converge two industries, is devoted more space. The third part provides the construction of the analysis.

1.4.1 Arguments for a Single Innovation Case Study

A single case study is an appropriate design when the case is critical for testing a well-formulated theory or propositions (Yin 1984:42–44). The main objective of the present study is to test the definition of the industry in building a theory, the cyclical model of technological change formulated by Anderson (1988) and Anderson and Tushman (1990). Since the challenging task of this study requires the identification of technological changes in the industry in time perspective, the research period is long, around 60 years. Due to the long time perspective, the present study is a longitudinal and historical.

Since a single case study design is an appropriate research design for testing a model (or the definition of the industry within a model), the technological innovations and especially the float glass innovation in the flat glass industry were chosen as a target in this historical case study. The reasons for choosing these innovations and this industry will be described in more detail in the following two sections. To understand and recognize the nature of the float glass innovation and the flat glass industry and the changes in it one must have a profound understanding of manufacturing technologies and several markets. A single case study of long time perspective in two markets fits this purpose. The empirical research design and the choices made are elaborated in Chap. 3.

1.4.2   The Flat Glass Industry and Innovations (Especially Float Glass) as a Target

The technological innovations and especially float glass innovation suit well for the analysis and synthesis of the empirical phenomenon of the present research. First, several radical technological changes, i.e. the introduction of drawing machine, continuous casting, and float glass into the flat glass industry occurred in the research period of the present study. All the elements of Anderson and Tushman's (1990) model exist: technological discontinuity (variation), era of ferment (selection), emergence of dominant design (selection) and era of incremental change (retention). The empirical case has the characteristics needed to examine theory and the definition of the industry in it. Second, the float glass manufacturing process, which converged two industries, was introduced at the latter half of the research period thus letting us to keep the two industries separate long enough.

As late as the mid-1970s, the flat glass industry contained two different subindustries : the sheet glass and the plate glass industries (Persson 1969; Doyle 1979, see Fig. 1.4.) Plate glass was needed in more sophisticated applications such as mirrors, automobile windows and the large windows used for retail displays and architectural effects, where inhomogeneities and optical distortion were not acceptable.

Sheet glass was cheap and it was subject to inhomogeneities and optical distortion; it was suitable for ordinary windows used in construction. These two subindustries were quite different. The plate glass industry was much more concentrated. Manufacturers in this branch were much larger than those in the sheet glass subindustry. This difference was due to the high investment requirements of the plate glass branch. Every plate glass manufacturer also manufactured sheet glass. However, the investment costs and production capacities were much smaller in the sheet glass branch than in the plate glass branch. For these reasons it was possible for smaller, independent companies to produce sheet glass.

The flat glass industry is an interesting example of the impact of a technological change, since a major technological change (float glass developed by Pilkington, UK) took place in 1959 in the flat glass industry and radically restructured the whole industry in the 1960s and the 1970s all over the world, as can be seen from the enclosed two quotations:

"Pilkington's discovery of float (glass) forced other glass manufacturers in the 1960s to switch production techniques almost overnight and the high cost of installing new plant meant that only larger, cash rich organizations could survive. The result has been to increase the concentration of manufacturing power within a small group of companies operating out of the UK, France, Japan and the U.S." (Taylor 1979)

"For the next decade (the 1960s), most activity in float glass manufacture was centered around the licensing of the PB (Pilkington Brothers) technology by flat glass and by the development of techniques for making a float product which was ever thinner and thinner." (Edge 1984:714/4).

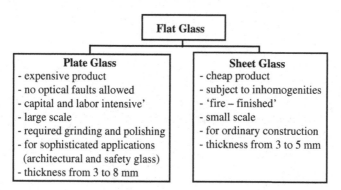

Fig. 1.4 The Division of Flat Glass in the 1950s

Float glass was quickly adopted in the capital-intensive plate glass industry. Float glass also accelerated globalization of this industry. Larger and less expensive production units were available and the number of Foreign Direct Investments (FDIs) increased. At the end of the 1960s, 10 years after its introduction, float glass was able to enter both the price competitive U.S. sheet glass market and the fragmented European, national sheet glass markets. Finally, in the mid-1970s this technological change, float glass, converged the flat glass industry into a single industry. Float glass was then as an innovations such as Levinthal (1998) described both rapid and slow.

Furthermore, the flat glass industry is a good laboratory for examining technology change since two separate markets (the U.S and Europe) can be studied. Both markets have their specific interests.

First, the U.S. market is attractive for many reasons. There were both plate and sheet glass producers in the U.S. market. In the late 1950s sheet glass production exceeded the plate glass production. In 1964 when float glass was introduced in the United States its production was 34 million sq. ft. The diffusion of float glass among plate glass producers was extremely rapid, but it took about 6–8 years before sheet glass producers, although the largest of them were also plate glass producers, adopted float glass. This is well demonstrated in Fig. 1.5. Edge (1984) also illustrated the similar curves for the production of sheet glass, plate glass and float glass for 1964 to 1980. As the U.S. flat glass industry was included in Anderson's (1988) study, it is possible to re-examine this industry in 1920–1980. Furthermore, Anderson (1988) and Anderson and Tushman (1990) did discover the convergence of the two subindustries, the plate glass and the window glass subindustries, although the empirical findings of this study indicate that they probably failed to grasp its nature.

Second, the European market is included in the present study since the main European plate glass (and sheet) manufacturers had played a remarkable role in the development of the flat glass market in the U.S. Also, the diffusion of the float glass process into the European market was different from that in the United

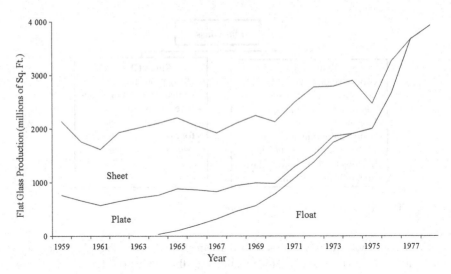

Fig. 1.5 Flat Glass Production in the United States, 1959–1978 (Data Source: Annual Current Industrial Reports, U.S. Department of Commerce)

States. The European producers were at first reluctant to obtain the float glass process, but later on some of them invested heavily in it.

1.4.3 The Construction of the Analysis

The construction of the analysis is described in Fig. 1.6. The units of analysis in the present research are the technological changes and especially the float glass innovation in the flat glass industry. Since the float glass innovation converged two industries it is on the highest level (Level I in Fig. 1.6) in the hierarchy of the analysis. The plate glass and sheet glass industries, two subindustries of the flat glass industry, form the core of the second level of the analysis. The third level of analysis deals with the processing industries important for different flat glass sub industries. Since the U.S. car industry had required since 1930s plate glass as the only raw material for safety glass it is included in the Fig. 1.6 with the dotted line. As one can see from Fig. 1.6 the flat glass industry (SIC 3211) is define by four-digit SIC code while the plate glass (SIC 32112), sheet glass (SIC 32111) and safety glass (SIC 32113) industries are defined by five-digit SIC codes. When introduced in the US market in the early 1960s float glass was classified under the plate glass code (SIC 32112). As we learnt earlier in 1977 float glass converged the SIC codes of plate glass and sheet glass under the code of plate glass (32112) The fourth level of the analysis contains the individual companies in both subindustries. Both of the sub and value added industries are studied in the United States and Europe.

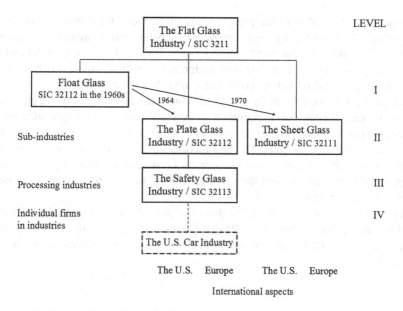

Fig. 1.6 The Construction of the Analysis

1.5 Outline of the Study

In Chap. 1, the linkages of this study to a wider research setting were described. The main purpose of the study and the research phenomenon were also introduced. Chapter 2 addresses the theoretical base, the concept of a cyclical model of technological change, proposed by Anderson (1988). This chapter is divided into two sections. First, the literature on which Anderson's concept of dominant design is based is briefly reviewed. Second, the cyclical model of technological change created by Anderson (1988) is described thoroughly.

The research methodology of the empirical study will be presented in Chap. 3. The methodology is based on the triangle—context, content and process—proposed by Pettigrew (1987b:5). Chapter 4 is devoted to the empirical part of this study. Here I use five-digit SIC codes in defining the industry. At the beginning of the chapter the flat glass products and technologies are briefly reviewed. Next, one closely related industry, the safety glass, which uses flat glass as its raw material, is described. This was done since this industry seems to have had a heavy impact on the development of the flat glass industry during the whole research period of 60 years. Then two subindustries, the plate glass (SIC 31112) and the sheet glass (31111) branches, in 1920–1960 are illustrated. The fourth section provides the empirical case descriptions of the evolution of the flat glass industry in 1960–1980 in two markets; 1) the United States and 2) Europe.

Chapter 5 has three sections. First, there is the description of Anderson's (1988) view of the development of the U.S. flat glass industry in 1960–1980 and the comments of the author of the present study on it. Next, three the key concepts, technological discontinuity, performance parameter and dominant design, of Anderson's model are discussed and redefined. The third section considers whether float glass emerged as a dominant design in two subindustries, the plate glass and sheet glass, and on two markets, the United States and Europe.

Chapter 6 provides the main theoretical contribution of this study. It proposes the new consequences of a dominant design observed in an application of a modified version of the concept of dominant design within the flat glass industry. This chapter has two sections. The first section illustrates the view of Anderson's (1988) model. It suggests two consequences or structural shifts (a gradual convergence of two industries and the concentration of the industry) which are caused by a dominant design.

In the final chapter, Chap. 6, the main findings of the present study are summarized. Also, discussions on the theoretical and managerial implications of this study are presented. Suggestions for future research are introduced at the end of the chapter.

References

Abernathy W (1978) The productivity dilemma (Second Printing). The Johns Hopkins Press Ltd, London

Abernathy W, Utterback J (1978) Patterns of industrial innovation. Technol Rev 7:40–47

Abernathy W, Clark KB (1985) Innovation: mapping the winds of creative destruction. Res Policy 14:3–22

Anderson P (1988) On the nature of technological progress and industrial dynamics, Unpublished Ph.D. dissertation, Columbia University

Anderson P, Tushman LM (1990) Technological discontinuities and dominant designs: a cyclical model of technological change. Adm Sci Q 35:604–633

Anderson P, Tushman LM (1991) Managing through cycles of technological change. Res Technol Manage 34:26–31

Arthur B (1998) Increasing returns and the new world of business. Harvard Bus Rev 74:100–110

Brittain JW, Freeman J (1980) Organizational proliferation and density dependent selection. In: Kimberly J, Miles RH (eds) The organizational life cycle. Jossey-Bass, San Francisco

Clark KB (1985) The interaction of design hierarchies and market concepts in technological evolution. Res Policy 14:235–251

Cusumano M, Gawer A (2002) The elements of platform leadership. MIT Sloan Manage Rev 43(3):51

David P, Greenstein S (1990) The economics of compatibility standards: an introduction to recent research. Econ Innov New Technol 1:3–42

Dosi G (1982) Technological paradigms and technological trajectories. Res Policy 11:147–162

Dosi G, Freeman C, Nelson R, Siverberg G, Soete L (eds) (1988) Technical change and economic theory. Pinters Publishers, London

Doyle PJ (ed) (1979) Glass making today. Portcullis Press Ltd, Redhill

Edge KC (1984) Section 11, flat glass manufacturing processes (Update). In: Tooley F (ed) Handbook of glass manufacture, 3rd edn, vols I, II, Books for the Glass Industry Division. Ashlee Publishing Co, New York, p 714/1–21

Freeman J (1982) Organizational life cycles and natural selection processes. In Staw BM, Cummings LL (eds) Research in organizational behavior, vol 4. JAI Press, Greenwich, pp 1–32

Foster R (1986) Innovation: the attacker's advantage. Summit Books, New York

Grønhaug K (1994) Lectures in the course: managing the doctoral thesis research, Järvenpää, September

Hannan MT, Freeman JH (1977) Population ecology of organizations. Am J Sociol 82(5):929–964

Hannan MT, Freeman JH (1989) Organizational ecology. Harvard University Press, MA

Katz M, Shapiro C (1992) Product introduction with network externalities. J Ind Econ 40:55–83

Levinthal D (1998) The slow pace of rapid technological change: gradualism and punctuation in technological change. Ind Corp Change 7(2):217–248

Liebowitz S (2002) Re-thinking the network economy. AMACOM, New York

Lumsden CJ, Singh JV (1990) The dynamics of organizational speciation. In: Singh JV (ed) Organizational evolution: new directions sage. Newbury Park, CA, pp 145–163

McGrath R, MacMillan IC, Tushman LM (1992) The Role of executive team actions in shaping dominant designs: towards the strategic shaping of technological progress. Strategic Manage J 13:137–161

McKelvey B (1982) Organizational systematics: taxonomy, classification, evolution. University of California Press, CA

Meyer M, Lehnerd A (1997) The power of product platforms. Free Press, New York

Nelson RR, Winter SG (1982) An evolutionary theory of economic change. The Belknap Press of Harvard University Press, Cambridge

Pavitt K (1984) Sectoral patterns of technical change: towards a taxonomy and a theory. Res Policy 13(6):343–373

Persson R (1969) Flat glass technology. Butterworths, London

Pettigrew AM (1987b) Introduction: researching strategic change. In: Pettigrew A (ed) The management of strategic change. Basil Blackwell, Oxford, pp 1–13

Porter M (1983) Technological dimension of competitive strategy. In: Rosenbloom RS (ed) Research on innovation, management and policy, vol 1. Jai Press, Greenwich, pp. 1–33

Reinganum JF (1981) On the diffusion of new technology: a game theoretic approach. Rev Econ Stud XLVIII:395–405

Romanelli E (1989) Organization birth and population variety: a community perspective on origins. Research in Organizational Behavior, vol 11. JAI Press, Greenwich, pp. 211–246

Rosenkopf L, Tushman LM (1994) The coevolution of technology and organization. In: Baum J, Singh J (eds) Evolutionary dynamics of organization. Oxford University Press, Oxford, pp 403–424

Sahal D (1981) Patterns of technological innovations. Addison-Wesley, New York

Schumpeter J (1934) The theory of economic development. Harvard University Press, MA

Schumpeter J (1942) Capitalism, socialism, and democracy. Harper and Brothers, New York

Suárez FF (2004) Dominant designs and the survival of firms. Res Policy 33:271–386

Suárez FF, Utterback J (1995) Dominant designs and the survival of firms. Strateg Manag J 16:415–430

Supple B (ed) (1977) Essay in british business history, Clarendon Press, Oxford

Taylor A (1979) Pilkington out to crack new markets. Financial Times, February 12

Teece DJ (1986) Profiting from technological innovation: implications for integration, collaboration, licensing and public policy. Res Policy 15:285–305

Teece DJ (1988) Technological change and the nature of the firm. In: Dosi G et al (eds) Technical change and economic theory. Pinter Publisher, London, pp 295–308

Thomson J (1967) Organizations in action. McGraw-Hill, New York

Tushman LM, Anderson P (1986) Technological discontinuities and organizational environments. Adm Sci Q 31:439–465

Tushman LM, Anderson P (1987) Technological discontinuities and organizational environ-
    ments. In: Pettigrew A (ed) The management of strategic change. Basil Blackwell, Oxford,
    pp 89–122
Tushman ML, Murmann JP (1998) Dominant designs, technology cycles and organizational
    outcomes. Res Organ Behav 20:231–266
Tushman LM, Romanelli E (1985) Organizational evolution: a metamorphosis model of
    convergence and reorientation. Res Organ Behav 7:171–222
Tushman LM, Rosenkopf L (1992) Organizational determinants of technological Change: toward
    a sociology of technological evolution. Research in organizational behavior. JAI Press,
    Greenwich, pp 311–347
Tushman LM, Newman W, Romanelli E (1986) Convergence and upheaval: managing the
    unsteady pace of organization evolution. Calif Manage Rev 29(1):29–44
Utterback J (1994) Mastering the dynamics of innovation how companies can seize opportunities
    in the face of technological change. Harvard Business School Press, MA
Utterback J, Abernathy W (1975) A dynamic model of process and product innovation. Omega
    33:639–656
Utterback J, Suárez FF (1993) Innovation, competition, and industry structure. Res Policy
    22:1–21
Van de Ven A, Garud R (1994) The coevolution of technical and institutional events in the
    development of an innovation organization. In: Baum J, Singh J (eds) Voluntary dynamics
    of organization. Oxford University Press, Oxford, pp 425–443
Yin RK (1984) Case study research, design and methods. Sage Publications, Newbury Park

Chapter 2
Technological Change: Dominant Design Approach

Abstract The cyclical model of technological change or dominant design model is based on the earlier dynamic models of technological change. These models such as product and process innovations, the transilience map, technological guideposts and creative symbiosis, technology s-curves plus evolutionary models are discussed first. Then the dominant design model and its elements, era of incremental change, era of ferment, technological change and dominant design are illustrated.

Keywords Dominant design model · Era of ferment · Technological change

This chapter which treats the theoretical background of the present study concerns the concept of the cyclical model of technological change proposed by Anderson (1988). This chapter also introduces dynamic models of technological change pertinent for Anderson's study. Moreover this literature review is essential in the re-examinations and re-testing of the cyclical model of technological change carried out in Chap. 5.

2.1 Anderson's Cyclical Model of Technological Change

This section discusses the cyclical model of technological change proposed by Anderson (1988). First, a concise review is devoted to the dynamic models of technological change on which Anderson (1988) based his model. Second, Anderson's model is discussed in detail. This sub-section also includes the definition of the concepts essential to the present study. The definitions for technology, innovation etc. are the same as those of Tushman and Anderson (1986), Anderson (1988) and Anderson and Tushman (1990, 1991).

O. Uusitalo, *Float Glass Innovation in the Flat Glass Industry*,
SpringerBriefs in Applied Sciences and Technology,
DOI: 10.1007/978-3-319-06829-9_2, © The Author(s) 2014

2.1.1 Dynamic Models of Technological Cycles

In the development of his dynamic model of technological cycles, dominant design, Anderson (1988) reviewed four dynamic models of technological evolution. Since Anderson had adopted certain features in his own model of each, all the models are briefly described. The first model, Abernathy and Utterback, is mainly based on three articles: Utterback and Abernathy (1975), Abernathy and Utterback (1978), Abernathy (1978) and Abernathy and Clark (1985). Three other models are those of Sahal (1981), Foster (1986) and Nelson and Winter (1982). Below is a brief review of these models plus Dosi's (1982) typology. Neither the above mentioned models nor Anderson's choices (i.e. the ideas from the above models included or excluded in/from his own model) are commented on or evaluated in the present study.

The models above are included in the present study because Anderson (1988) obtained several fundamental ideas from them. First, technology progresses in an irregular pattern, characterized sometimes by incremental and sometimes by major breakthroughs, which advance a technology past its previous limits. Second, these epochal innovations are followed by a period of de-maturity, experimentation and flexibility in R&D. Third, a dominant design or technological paradigm, or technical guidepost emerges, thus establishing the direction of future change along few natural trajectories, until some event triggers another change (Anderson, 1988:40). Anderson created his model of dominant design from the following reasons: "It (the model) allows us to understand different types of innovation in different industries during different periods using a common framework that lets us compare one to another" (Anderson, 1988:42).

2.1.1.1 Abernathy and Utterback

The Utterback and Abernathy (1975)/Abernathy and Utterbak (1978) model presumes that a major product innovation inaugurates the cycle. The output rate stimulated by minor innovations appears before the technology stimulated by major process innovation emerges. Abernathy and Utterback (1978) divided the Stage of Development into three patterns and named them the fluid, transitional and specific patterns. In the infancy of an industry (i.e. in the fluid pattern) firms are small, informal, flexible and entrepreneurial. The product line diversity arises from fundamental differences in technology. The competitive advantage is obtained by a technical superiority which allows a higher margin for their products.

According to Abernathy and Utterback, although many observers emphasize radical product innovations, process and incremental innovation may have equal and even greater commercial importance. The design usually creates a number of proven concepts and is seldom an advance in the state-of-the-art. The emergency of dominant design alters the pattern of technological change. The key technological development after a dominant design is cost reduction due to learning.

In 1985 Abernathy and Clark (1985) presented a new framework for analyzing the competitive implications of innovation. The framework is based on the concept of transilience, which means the capacity of an innovation to influence the established systems of production and marketing (Abernathy and Clark 1985:3). In the transilience map two diagonals—the extent to which an innovation disrupts existing production and operation competencies and the extent to which it disrupts existing marketing linkages—divide innovation into four categories: Architectural, Niche Creation, Regular and Revolutionary (ibid., p. 7).

Abernathy and Clark (1985) use the term epochal innovation in order to distinguish a radical innovation from an incremental innovation, which reinforces the existing tendencies in process development. An epochal innovation occurs with the introduction of an approach that cannot be produced effectively with the existing production processes. Clark (1985:112) says: "The essential issue in defining 'epochalness' is how a given innovation affects existing capital equipment, labor skills, materials and components, management expertise, and organizational capabilities." Niche creation and regular innovations are incremental and architectural and revolutionary ones are epochal.

Technological evolution in this transilience map-model goes counterclockwise from the northeast quadrant to the southeast quadrant (i.e. architectural, niche creation, regular and revolutionary). An architectural innovation opens market for new product classes. With the emergence of a dominant design the industry moves via the niche creation phase to the regular innovation phase. Abernathy and Clark (1985) cites Kuhn (1972), who suggests that the advancement of science is characterized by long periods of regular development, punctuated by periods of revolution. Historical evidence suggests that a similar pattern characterizes the development of technology, i.e. an epochal innovation can start the second round (referred to as industry de-maturity, Clark 1985:112–115) in the transilience map (Abernathy and Clark 1985:14).

Anderson does not use the three ideas included in these Utterback and Abernathy (1975), Abernathy and Utterback (1978) and Abernathy and Clark (1985) frameworks:

1. Anderson does not presume that product innovations typically inaugurate or dominate each cycle.
2. He abandons the notion that epochal innovations need not improve key performance parameters.
3. Anderson also abandons the presumed evolutionary path of technologies in counter clock wise from architectural innovation via niche creation and regular innovations to revolutionary one.

2.1.1.2 Sahal (1981)

Anderson mentioned that Sahal's model extends to concerns that are outside the scope of his work, for instance how breakthrough innovations arise, how long-wave economic cycles are linked to secular trends in national technological development, and how technologies diffuse in space and time. Anderson focuses on two key ideas: technological guideposts and creative symbiosis. The notion of technological guideposts can often be seen in one or two early models of technology which stand out above all others in the history of an industry. Their design becomes the basis for many innovations via a process of gradual evolution. The designs left imprints on a whole series of observed progresses in technology. Sahal uses Farmall and Fordson tractors and DC-3 aircraft as examples.

According to Anderson (1988), the concept of technological guidepost remains that of dominant design, but the emphasis is on how a standard imprints subsequent designs, not on the role of the standard in shifting an industry toward volume production (Anderson 1988:30). Anderson continues (ibid., p. 31), "a dominant design is a standard configuration that influences and constrains subsequent generations of a technology; it is not assumed here (in Sahal's work) that its dominance is based on its production in volume." Creative symbiosis is a situation where two individual core technologies have approached their limits and are then integrated to simplify the overall structure, thereby circumventing the limits to its evolution. In Sahal's work the empirical evidence comes from the farm tractor.

2.1.1.3 Foster's S-curves

According to Foster (1986), most of the managers of companies that enjoy transitory success assume that tomorrow will be more or less like today. Significant changes are unlikely, unpredictable, and they in any case come slowly. Foster's S-curve is a graph of the relationship between the effort put into improving a product or a process and the results one gets back from that particular investment. At the beginning, as money is put into a new product or a new process development, progress is very slow. Then something happens as more learning and the key knowledge necessary to make advances is put in place. Finally, as more money is put into development of the product or the process, it becomes more and more difficult and expensive to make technical progress. The S-curve sets the limit to a particular technology.

The quotation from Foster (1986:34) explains the importance of an S-curve. "If you are at the limit, no matter how hard you try you cannot make progress. As you approach limits, the costs of making progress accelerates dramatically. Therefore, knowing the limit is crucial for a company if it is to anticipate change or at least stop pouring money into something that can't be improved. The problem for most companies is that they never know their limits. They do not systematically seek the one beacon in the night storm that will tell them just how far they can improve their products and processes." According to Foster, S-curves almost always come in pairs. The gap (or the movement) between the pair of S-curves represents a technological discontinuity—a point when one technology replaces another. Rarely does a single technology meet all the customers' requirements and the many technologies compete with each other.

2.1.1.4 Nelson and Winter (1982) Model

Nelson and Winter (1982) do not take technology as given, which is opposite to the position of microeconomic theory. Traditional economics has also utilized maximizing assumptions which lead to equilibria. In traditional economics the equilibria are the pillars and the explanation of economic change usually involves a move from one equilibrium to another in response to an exogenous driving force. On the contrary, the firms in the Nelson and Winter evolutionary theory are treated as though they were motivated by profit and engaged in a search for ways to improve profits, although their actions are not assumed to be profit maximizing over well defined and exogenously given choice sets. This evolutionary theory emphasizes the fact that the most profitable firms drive the less profitable ones out of the industry; however, Nelson and Winter do not focus in their analysis on the hypothetical states of industry equilibrium in which all poorly performing companies no longer operate and the well performing ones are at their desired size. Nelson and Winter (1982:4).

Nelson and Winter (1982) stress the constant action of the natural selection mechanism over time, which winnows out less profitable organizations. Firms are not subject to a maximizing calculus in this model; instead, they have certain capabilities and rules which are modified over time, and which confer selection advantages or disadvantages. A key source of variation, which provides the material for selection to operate, is technological change. They say: "The core concern of evolutionary theory is with the dynamic process by which firm behavior patterns and market outcomes are jointly determined over time. Through the joint action of search and selection, the firms evolve over time, with the condition of the industry in each period bearing the seeds of its condition in the following period. It is precisely in the characterization of the transition from one period to the next that the main theoretical commitments of evolutionary theory have direct application." (ibid., 1982:18).

The influence of the work of Schumpeter in the Nelson and Winter evolutionary theory has been pervasive (ibid., 1982:39). Schumpeter was the first to develop a theory of economic progress that assigned a key role to technological change (Schumpeter 1934; Schumpeter 1942). The crucial matter in Schumpeter's analysis is the 'process of creative destruction.' If creative destruction is the driving force behind economic progress, then a theory of evolutionary economic change must take into account how technologies change over time and how industries adjust to technical advances. Nelson and Winter model this phenomenon.

According to Nelson and Winter, a technology has both economic parameters (which will affect its cost and/or the demand) and technological parameters (such as size, chemical composition) (1982:248). The R&D decision maker does not in general know the economic attributes (for instance how much the new technology will reduce prices or create demand). He knows at least some of the technical attributes. To find out more about the technical and economic parameters, he or she conducts a search; because of the uncertainty surrounding these parameters it is not clear that the firm will locate or organize the best possible technology as a result.

According to Nelson and Winter, most technological regimes are characterized by 'natural trajectories,' paths which technicians believe to be feasible and worth attempting (1982:258). These trajectories compose a few routes which direct R&D, since they seem to be the most logical and promising ways to lower costs or to create attractive new products.

2.1.1.5 Dosi

Anderson emphasis the role of the dominant design in establishing trajectories and especially argues that dominant designs arise because organizations seek certainty in their search patterns (Anderson 1988:39). This idea of dominant design is according to Anderson akin to Dosi's (1982) argument that technological trajectories are shaped by 'technological paradigms.' Dosi defines a technological paradigm as "a model and pattern of solution of selected technological problems based on selected principles derived from natural sciences and on selected material technologies" (Dosi 1982:152). A technological trajectory is a pattern of "normal problem solving activity... on the ground of a technological paradigm" (ibid., 1982:152). According to Dosi, technological paradigms emerge for a variety of reasons, often institutional and/or political rather than technological (ibid., 1982:155). Dosi stresses that there is probably more uncertainty at the early stages of an industry's history and more firms are gambling on different technological paradigms; "competition does not occur between the 'new' technology and the 'old' one which it tends to substitute but also among alternative 'new' technological approaches" (ibid., 1982:155). Anderson refers technological paradigm to dominant design and technological trajectory to the incremental change process.

2.1.2 The Elements of Anderson's Technological Cycle

This section describes the elements of Anderson's technological change. His definitions of the key concepts are included in this section as was mentioned earlier. Technology is defined according to Rosenberg (1972) as those tools, devices, and knowledge that mediate between inputs and outputs (process technology) and/or that create new products or services (product technology). The social, managerial and technologies other than product and equipment technologies are left out of Anderson's work (Anderson 1988:14) and also of the present study. Moreover, in this study technology refers only to process technology. Anderson defines technological change as an event which lowers the production cost or improves the performance of an industry's output by substituting a technically superior product or process for its predecessor. Anderson refers to Nicholson (1987) and says that the movements along the production functions (isoquants) in either dimension are not technological changes.

Technological innovation is defined as the first commercial introduction of a product or a process in an industry when that introduction constitutes a technological change as defined above (Anderson 1988:17). According to Anderson, his dominant design model relies heavily on the concept of technological discontinuity (1988:17). The discontinuity between major and minor innovations is common, but the basis of the distinction often varies from study to study. Arrow (1962), for instance, defines a major innovation as one which pulls the prices of a product below its previous perfectly competitive price; a minor innovation is one which does not. According to Anderson, Arrow's definition ignores the performance dimension and does not differentiate innovations which affect the competitive price a great deal from those which reduce it by a small amount (Anderson 1988:18).

According to Anderson (1988), to identify a technological discontinuity one has to be able to track progress in key performance parameters over time. In the flat glass industry (Anderson, 1988:105–106) output is essentially a commodity, and since scale economies are important, the key determinant (or performance parameters/dimensions of merit) of production cost is the output per unit of time of the most productive process equipment in existence. In plate and window glass, this was the square feet per hour capacity of a flat glass forming machine.

It is also evident that technological discontinuities are not all alike. Tushman and Anderson (1986) characterized technological discontinuities as competence-enhancing or competence-destroying. On the one hand, competence-enhancing discontinuities significantly advance the state of the art yet build on, or permit the transfer of, existing know-how and knowledge. Competence-destroying discontinuities, on the other hand, significantly advance the technological frontier, but with a knowledge, skill and competence base that is inconsistent with prior know-how. Competence-destroying discontinuities are so fundamentally different from previously dominant technologies that the skills and the knowledge base required to operate the core technology shift. The older technology quite seldom vanishes quietly; competition between old and new technologies is fierce (Foster 1986). That kind of major changes in skills, competence, and production processes are associated with major changes in the distribution of power and control within firms and industries (Chandler 1977).

In Anderson's study (1988:21) the industry is defined by Standard Industrial Classification (SIC) codes. In the statistics of this study, the industry is defined by SIC codes or by Customs Co-operation Council Nomenclature (CCCN) codes. Since these definitions set strict borders for industries, other types of qualitative measures are used to determine the links between the industries. This is an essential factor in this study since the value added products have a significant impact on the diffusion of new technology for the base product.

In Anderson's model a discontinuous technological innovation (see Technological Discontinuity 1 in Fig. 2.1) initiates a series of cycles which create a pattern of technological change in an industry. The usual course of technical progress in an industry consists of long periods of incremental change. A technological discontinuity arises, interrupts the era of incremental change and starts

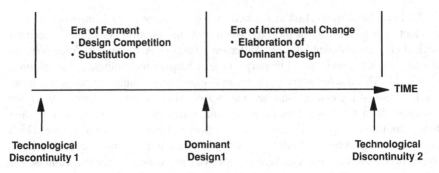

Fig. 2.1 The technology cycle (Reproduced from Anderson and Tushman 1990)

one of ferment at random intervals. Technological discontinuities initiate substantial technological rivalry between alternative regimes. Social, political and organizational dynamics select single industry standards or dominant design from among technological opportunities.

The emergence of dominant design ends the era of ferment and starts another era of incremental change, which is again interrupted by a new technological breakthrough. Anderson and Tushman argue (1990:606) that a breakthrough innovation inaugurates an era of ferment in which competition among variations of the original breakthrough culminates in the selection of a single dominant configuration of the new technology. Successful variations are preserved by the incremental evolution of this standard design until a new discontinuous technological change initiates a new cycle of variation, selection and retention. The cyclical model of technological change with its elements (i.e. era of ferment, dominant design and era of incremental change) is illustrated in Fig. 2.1. The elements are discussed below in greater depth.

Era of Ferment. The era of ferment is characterized by two processes. The first is technological substitution; the innovative product or process replaces the prior technical regime of the industry. The second is design competition. This period of substantial product class variation and, in turn, uncertainty, ends with the emergence of a dominant design. While dominant designs are critical at the process or product class level, for a given company, betting on a particular industry standard or technology involves substantial risk (Anderson and Tushman, 1990). An operational definition of the length of the era of ferment is that period starting with the year in which a discontinuous innovation commercially appeared on the market and including the first of the three consecutive years in which a design is applied in 50 % share of the new installations (Anderson 1988:108) see also the operational definition of dominant design below).

Dominant design. Finally, the era of ferment ends with the appearance and the predominance of a single design. Ultimately, one basic implementation of the technical breakthrough predominates and eliminates rival designs, for both technical and organizational reasons (Anderson 1988:44). On the technical side, a dominant design combines small advances into an effective combination. As the dominant design becomes the focus of R&D effort, one improvement leads to another and the dominant design becomes technically superior to its rivals. When the production volumes become higher learning by both doing and using takes place, thus lowering the production costs below those of competing designs (Anderson 1988:44). On the organizational side, the companies are driven to seek and adopt a technological standard. They try to avoid uncertainty (March and Simon 1958), and seek rationality, particularly in the technical core (Thompson 1967). The dominance of a substitute product, a substitute process, or a dominant design is a function of technological, market, legal and social factors that cannot be fully known in advance (Anderson and Tushman 1990). The operational definition (Anderson 1988:107) for a dominant design is a single configuration or narrow range of configurations that accounted for at least 50 % of the new process installations or new products in at least three consecutive years following a discontinuity.

Era of Incremental change. The emergence of a dominant design changes the competitive landscape (Utterback and Abernathy 1975). After a dominant design emerges, technological progress is driven by numerous incremental innovations (Myers and Marquis 1969 in Anderson and Tushman 1990). Variation now takes the form of elaborating the retained dominant design instead of challenging the industry standard with new, rival architectures. The focus of competition shifts from higher performance to lower cost and to differentiation via minor design variations and strategic positioning tactics (Porter 1985). The era of incremental change is characterized by incremental, competence-enhancing, puzzle-solving actions of many organizations that are learning by doing (Anderson and Tushman 1990:606). The cyclical model of technological change or the technology cycle is illustrated in Fig. 2.1.

2.2 Summary of the Literature

This section briefly summarizes the most important aspects of the concept, which will be evaluated. First, the operationalization (i.e. the definitions of the concepts) of the model is discussed. As was mentioned earlier (Sect. 1.2, p. 11), the definition of performance parameter in Anderson's (1988) study seemed to be weak. Anderson's performance parameter for sheet glass, plate glass and cement manufacturers measures the efficiency of an organization instead of its effectiveness (Pfeffer and Salancik 1978). Moreover, for some reason Anderson did not regard float glass as the dominant design although float glass is regarded as one of the most elegant innovations of this century (Caulkin 1987; Arbose 1986). It seems that operationalization of the model is the weakest point in Anderson's model and in his testing of the hypotheses. These will be discussed thoroughly in Chap. 5.

References

Abernathy W (1978) The productivity dilemma, second printing. The Johns Hopkins Press Ltd, London

Abernathy W, Utterback J (1978) Patterns of industrial innovation. Technol Rev 7:40–47

Abernathy W, Clark KB (1985) Innovation: mapping the winds of creative destruction. Res Policy 14:3–22

Anderson P (1988) On the nature of technological progress and industrial dynamics, Unpublished PhD dissertation, Columbia University

Anderson P, Tushman LM (1990) Technological discontinuities and dominant designs: a cyclical model of technological change. Adm Sci Q 35:604–633

Anderson P, Tushman LM (1991) Managing through cycles of technological change. Res Technol Manage 34:26–31

Arbose J (1986) Why once-dingy Pilkington has now that certain sparkle. Int Manage 41(8):44–45, 48

Arrow K (1962) The economic implications of learning by doing. Rev Econ Stud 29:155–173

Chandler A (1977) The visible hand: the managerial revolution in American business. Belknap Press, Cambridge

Caulkin S (1987) Pilkington after BTR. Management Today, June:43-49, 127, 129

Clark KB (1985) The interaction of design hierarchies and market concepts in technological evolution. Res Policy 14:235–251

Dosi G (1982) Technological paradigms and technological trajectories. Res Policy 11:147–162

Foster R (1986) Innovation: the attacker's advantage. Summit Books, New York

Kuhn T (1972) The structure of scientific revolution. University of Chicago Press, Chicago

March J, Simon H (1958) Organizations. Wiley, New York

Myers S, Marquis D (1969) Successful industrial innovations. National Science Foundation, Washington

Nicholson W (1987) Intermediate microeconomics and its application, 4th edn. The Dryden Press, New York

Nelson RR, Winter SG (1982) An evolutionary theory of economic change. The Belknap Press of Harvard University Press, Cambridge

Porter M (1985) Competitive advantage. Free Press, New York

Pfeffer J, Salancik G (1978) The external control of organizations: the resource dependence perspective. Harper & Row, New York

Rosenberg N (1972) Technology and american economic growth. ME Sharpe, New York

Sahal D (1981) Patterns of technological innovations. Addison-Wesley, New York

Schumpeter J (1934) The theory of economic development. Harvard University Press, Cambridge

Schumpeter J (1942) Capitalism, socialism, and democracy. Harper Brothers, New York

Thomson J (1967) Organizations in action. McGraw-Hill, New York

Tushman LM, Anderson P (1986) Technological discontinuities and organizational environments. Adm Sci Q 31:439–465

Utterback J, Abernathy W (1975) A dynamic model of process and product innovation. Omega 33:639–656

Chapter 3
Research Methodology

Abstract This chapter gives arguments for the methodological choices of the empirical study. Because this study re-test based on the empirical material from the flat glass industry the cyclical model of technological change it crucial that empirical study is well conducted. The research method is a comprehensive qualitative longitudinal and contextual case study. The study period of the particular technological change, float glass, starts almost four decades before the introduction of float glass and lasts two decades after it. This selection guarantees that the context is understood. The study uses rich case material collected from interviews of industrial participant, business histories, management journals etc. The quantitative statistics are also used. The validity of data analysis has been increased by pattern matching, pattern recognition, seeing evidence through multiple lenses and data triangulation.

Keywords Comprehensive case study · Contextual study

The purpose of this chapter is to give arguments for the methodological choices of the empirical study. Since this study re-examines and re-test based on the empirical material from the flat glass industry the cyclical model of technological change introduced by Anderson (1988) it crucial that empirical study in conducted well. The research method is a comprehensive qualitative longitudinal case study in the contrast to the quantitative method used by Anderson. The study uses rich case material collected from interviews of industrial participant, business histories, management journals etc. The quantitative statistics are also used. Special attention will also be given to the quality of the data.

3.1 Research Approach

The main objective of this study is to evaluate the definition of the industry in the development of the concept of cyclical model of technological change developed by Anderson (1988). The empirical purpose derived from the main objective is

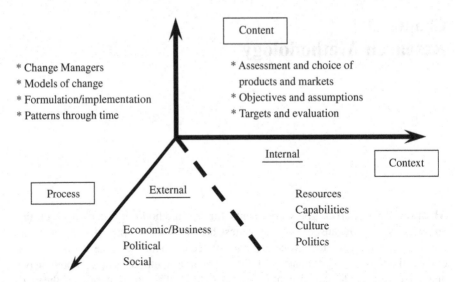

Fig. 3.1 Understanding strategic change: three essential dimensions (reproduced form Pettigrew and Whipp 1991)

divided in two parts. The first part is derived from the first framework (Anderson 1988). An attempt is made to capture the effects of a technological change (and the diffusion of new technology) on the industry structure as broadly as possible from the introduction of the new technology.

The main objective and the empirical purpose require a processual research approach, which includes the time perspective. This kind of research approach is also suggested by Pettigrew and Whipp (1991). In his study of organizational development in the Imperial Chemical Industries, Pettigrew (1985) argues that to understand a change we have to study it as a continuing process in the context in which it appears and he encourages us to adopt a contextual and historical perspective on processes of change, whatever the content of the change might be. According to Pettigrew and Whipp (1991) it is essential to explore these three basic concepts and their interconnections through time (see Fig. 3.1). Although the present research does not tackle strategic change at the firm level directly, this conceptualization helps in understanding the empirical data of this study.

Since the research on the effects of a technological change on the industry structures requires for instance the recognition of an innovation (or introduction of a new technology) in the industry, the calculation of the new manufacturing installations and withdrawals from the industry, etc., it is extremely important to know the industry concerned and the reasons for withdrawing from it. Since industries and reasons for withdrawals change all the time the competition (or internal and external contexts) is (are) best appreciated in a two dimensional way (Pettigrew and Whipp 1991). The dimensions are the levels at which competition operates and the element of time as indicated in Fig. 3.2. In the vertical axis there

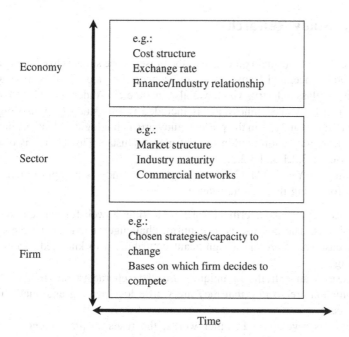

Fig. 3.2 Competition: three levels across time (reproduced form Pettigrew and Whipp 1991)

are the three major levels with associated characteristics and measures. The competitive performance of a firm is therefore based on the recognition that firms compete not merely against one another but at the same time within the sectoral (industrial) and national /international structures and relationships.

Pettigrew and Whipp (1991:28) further state that as the horizontal axis suggests, the sectoral and national conditions in which a firm operates and competes are usually unstable. They also mention that firms have different abilities to perceive those changes and to react to them. The categorization (see Fig. 3.2) is extremely helpful since it identifies many important aspects which were changing during the research period. This conceptualization is used in the operationalization of the present research.

In brief, the research approach in this study is longitudinal, historical and contextual. The aim in this research is to choose "a video camera view instead of a snap shot" as Pettigrew and Whipp's (1991) put it. I apply Pettigrew's (1990) longitudinal strategy. The research approach used in this study has allowed the researcher to look at the research phenomenon from the conceptual perspective and then from the empirical data and vice versa in the way of an abductive study (Dubois and Gadde 2002). This approach is similar to that of Cipolla's (1991) (in Lundgren 1991:75): "The aim of a research is not to twist facts to prove a theory, but rather to adapt theory to provide a better account of the facts. In the process of scientific inquiries into unique historical events, it must therefore be perpetual feedback between the formulation of problems and the process of gathering evidence."

3.2 Case Study Research

As was mentioned, the main objective of the present case study is to evaluate and re-examine one conceptual framework. This concept or model will be re-examined within the flat glass industry, which was also included in Anderson's (1988) original research. The float glass innovation is included in this study (see also Fig. 1.6). Thus, the units of analysis in the present study are technological changes, including the float glass innovation, within the flat glass industry. This choice is discussed further in Sects. 3.2.1 and 3.2.2.

According to Yin (1984:42–44), a single case study is an appropriate design under the following three circumstances:

1. The case represents a critical case in testing a well-formulated theory or propositions. The purpose is to confirm, challenge or extend the theory. The single case can represent a significant contribution to knowledge and theory building.
2. The case is an extreme or unique case. The characteristics of the research phenomenon are so rare that they are worth documenting and analyzing in a single case.
3. The case is revelatory. In other words, the research phenomenon has been previously inaccessible to scientific investigation.

Since the present case (the float glass innovation) can be regarded as a critical case the choice of only one innovation is appropriate. The present case is also in accord with Lilja (1983) who argues that a strategic (a synonym to critical) case is appropriate when there are alternative theoretical and methodological grounds according to which the prevailing hypothesis can be put under suspect (1983:351). Hagg and Hedlund (1978) (in Zander 1991:90) argue similarly that case studies can have important roles in the 'testing' of hypotheses. The purpose of the empirical research derived from the main objective is to find out and understand what kind of effects a radical technological change can have on industry structures and the complexities also involved in the innovation and diffusion processes of a product. A single case study of an innovation in the flat glass industry is suitable for the requirements of both the theoretical and empirical objectives.

A single case study design has certain advantages compared with multiple cases. The most important is the depth of the analysis, both in terms of the number of factors studied and sources of information used (Yin 1984). A single case analysis is the best way to get a holistic picture and understanding of the research problem. According to Siggelkow (2007) case studies may be used as illustration in the context of making a conceptual contribution. Patton (1990:95) argues along the same lines as Yin when he says that "qualitative inquiry is highly appropriate in studying process because depicting process requires detailed description." Since the units of analysis in the present single case study are innovations within two subindustries, the plate and the sheet, of the flat glass industry, and within firms in two different markets (see Fig. 1.6), the final design of this study is an "embedded, multiple case study design" (Yin 1984:41).

Like any research approach, the case study has its limitations. One of the biggest concerns has been the lack of rigor of case study research. The methods of analysis are not well-formulated in the use of qualitative data (Miles 1979). The case study research is very time-consuming and results in massive documents, which require special skills to handle (Yin 1984). Another limitation of case studies is that they provide very little basis for scientific generalization.

Porter (1980) points out that in an industry analysis there are important benefits in getting an overview of the industry first, and only then focusing on the specifics. According to him, experience has shown that a broad understanding can help the researcher to spot important items of data when studying sources and organize data more effectively as they are collected. Porter (1981: 449–451) also stresses the value of in-depth industry histories in understanding industry environments and identifying firms' strategic interactions on a longitudinal basis.

3.2.1 Choice of the Case Innovations

There are four reasons for the choice of the float glass innovation in the flat glass industry as the research target. First, float glass, the invention, was so radical that it fulfills the requirements of the technological change needed to test Anderson's (1988) model. Second, Anderson (1988), whose model is also going to be tested, includes the U.S. flat glass industry in his own research for the time period 1607–1980 (Anderson 1988:337) and thus float glass is also included. This allows us to re-examine the flat glass industry (thus the float glass innovation) in that particular market (or in the United States) in 1960–1990.

Third, the float glass innovation and thus the flat glass industry is also empirically interesting for three reasons: (a) The technological change, i.e. float glass, gradually converted the two subindustries, the plate glass and the sheet glass branches, into one industry; (b) The diffusion of float glass into two markets and different subindustries was dissimilar; (c) This discrepancy in diffusions created vast uncertainty, especially among smaller sheet glass manufacturers, since some major flat glass producers were betting on both the new float glass technology and the existing sheet glass technology.

Based on the theoretical frameworks used in this study, the empirical material from the time period 1920–1980 is illustrated in chronological order in two sections. The first part, the era of 1920–1960, is described without separating markets. The second part, the time period of 1960–1980, is illustrated in two markets. This market separations is made because of the different market structures and the different diffusion of float glass in these two markets. The reasons for the choice of the two markets, the United States and Europe, are discussed in the following paragraphs. As will be seen below, these reasons are interlinked (i.e. the choice of one market has an effect on the choice of another).

The United States. The choice of the U.S. market is based on several reasons. First, Anderson (1988) includes the U.S. flat glass industry in his study. Second, there were both sheet and plate glass producers in the U.S. market which enable us to compare the diffusion of float glass into the plate glass and sheet glass industries. The diffusion of float glass among plate glass manufacturers was extremely rapid compared with that among sheet glass producers. Third, the flat glass industry in the U.S. is well documented.

Europe. There are also many reasons supporting the choice of Europe as a target market. First, the re-examination of the models with the float glass innovation requires knowledge of the origins of float glass. Since the float glass innovation, the unit of analysis in the present study, was developed in the UK, this market should be included in the present study. Second, due to the international nature of the flat glass industry it is difficult or nearly impossible to investigate the U.S. and European markets separately. For instance, flat glass technology since the early 1900s has been licensed or cross-licensed between the U.S. and European manufacturers.

After the Second World War the automobile and construction industries were the two main industries which needed flat glass. For instance in the U.S., the auto industry, one large user of safety glass, had typically consumed 35–45 % percent of the flat glass produced in the 1950s. In the 1940s and the 1950s safety glass was mainly made from plate glass. The safety glass industry, which supplied products to the growing auto industry, emerged on the U.S. and the European market in the 1920s and the 1930s. The multi glass unit (MGU) industry, which supplied heat and sound insulation glass to the construction industry, emerged in these two markets in the 1930s. The safety glass industry is included in the analysis while the MGU industry is excluded from the present study.

From the empirical research material, the special flat glass types (rolled, patterned, wired, etc.) and mirrors, which represent one type of value added product, are excluded from the analysis. Safety glasses are basically divided into two categories; toughened and laminated. Flat glass can be toughened in two ways: thermal strengthening (also called tempering) or chemical strengthening. Corning Glass in the U.S. introduced chemical strengthened glass for the auto industry in the mid-1960s. Because most glass is tempered thermally, chemical strengthening is excluded from the present study. Thus toughened glass means tempered glass.

3.2.2 Choice of Research Period

The period covered by the present research is 60 years (1920–1980). The technological breakthrough, float glass, was introduced in 1959. To see and understand the invention process, the research period should start much earlier than the day of the announcement of the innovation, i.e. float glass. Thus the year 1920 has been chosen as the starting point. However, when the flat glass manufacturing technologies are concerned the research period starts earlier, i.e. in the 1910s since the major innovations in the flat glass industry before float glass occurred also in this decades. The re-examination of Anderson's model, on the other hand, requires that the research period is long enough after the technological change, i.e. float glass. In Anderson's study the research period ended in 1980.

3.3 Operationalization and Data Collection

The operationalization depends on the needs of the study. I am interested in the emergence of a technological discontinuity or an innovation and the effects of this technological changes on industrial structure. Thus, it is first important to identify technological change and second, their possible effects on the industrial structure. My strategy in data collection followed that presented earlier by Porter (1980). First, I had to get an overview of the industry before I focused on the specifics. The teaching case (Uusitalo 1993) was a tool to achieve an overview. The case was written together with industrial experts (see more Uusitalo 1995b).

The main objective of the present study is to test a theory, i.e. a cyclical model of technological change. This concept to be tested guided the operationalization of the research.

The rest of this section describes the operationalization of the concepts used in this study. Appendix 1 provides a summary of the operationalization. Process technology is a way to produce flat glass. Here we have to be able to identify the plate, sheet and float glass manufacturing processes. We also must be able to distinguish these manufacturing processes. This means that we have to be able to evaluate what natural sciences and engineering capabilities each process requires. Also, we must know what kind of inter- and intraorganizational matters were involved in the first phases (before the announcement) of the development of float glass process. In order to evaluate the condition of the sheet glass and plate glass manufacturing processes at the time float glass was invented, i.e. their novelty, we have to be able to identify their introduction time, the production capacities, the costs of the flat glass produced, the investment costs, the future potential offered, etc.

Technological innovation takes place at the moment the first square meter of float glass is sold in the market. This means that we must be able to identify the first square meters of float glass sold. We must also be able to identify the sub-industry in which float glass was invented. Technological discontinuity in this study has the same operationalization as technological innovation. The effect of technological change (i.e. float glass) is measured by performance parameter, which in Anderson's study is the production volume (square feet/h) of a flat glass manufacturing process. The redefined performance parameter in this study is the production costs of a manufacturing process for flat glass of a nominal thickness. The identification of production costs (or selling price) and the available thickness of float glass are crucial since the sheet glass markets were price-sensitive and the flat glass sold there was far thinner than the first available pieces of float glass. Float glass was a competence-destroying discontinuity for both plate glass and sheet glass manufacturers. Plate glass and sheet glass subindustries are defined according to five-digit Standard Industrial Classification (SIC) or Customs Co-operation Council Nomenclature (CCCN) codes.

In order to identify the emergence of float glass as a dominant design we first have to be able to verify whether float glass is within a single configuration or within its narrow range and second to calculate the market shares of each new flat glass manufacturing process installation. The first prerequisite is to understand what is meant by 'within narrow range.' It means that technology (the float glass process) in a new production line is based on the same technology which was defined as a technological change. In the present research the technology in a new float glass line is 'within narrow range of configuration' if the technology is licensed by Pilkington. This definition is supported by Suárez and Utterback (1995) who say the a notable increase in licensing activity over several years by a given firm or by a group of firms with products based on the same core technology is a better measure to determine a dominant design than, for instance, market share. Thus, a licensed float glass line is clearly within the narrow range of configurations. The second one requires the identification of new flat glass installations. The measurement of the era of ferment is based on the same method as the identification of the emergence of dominant design since the era of ferment is the time between the introduction of float glass and the first year of the three consecutive years in which the float glass manufacturing installations have a market share of over 50 %.

As mentioned earlier, Pettigrew and Whipp's two dimensional conceptualization of competition (see Fig. 3.2) was used in the analysis of the flat glass industry. Below are some reasons for taking this framework. The lower cost structure in Europe made it possible to import cheap sheet glass to the United States. The devaluations of foreign currencies also increased imports of sheet glass to the United States. Thus, we must be able to distinguish the effects of the foreign competition on the price sensitive U.S. sheet glass industry. At the sector/industry we must be able to recognize the effects of Pilkington's restrictive licensing policy on the emergence of float glass as the dominant design. In brief, sectoral /industry level changes were enormous. Float glass converged the two subindustries into one. This technological change also globalized the flat glass industry.

The above operationalization process and the choice of data sources were easier after the overview of the flat glass industry was acquired in the case writing process. Yin (1984:79) gives several sources of evidence for the data collection. In the case writing, documentation and a limited number of interviews were used. The purpose of both personal and telephone interviews was to make the overview as accurate and truthful as possible. During the case writing process understanding of the two subindustries, the role of value-added industries and different flat glass manufacturing processes increased substantially. Media were always present in inaugurations of new manufacturing process installations, which made it easy to identify either new sheet glass or float glass installations and the market shares of either sheet or plate glass installations. The plate glass industry was easier to follow since the last plate glass line built in the world started up in the United States in 1962.

The data required in the above operationalization of the present study were collected mainly from documentation, i.e. industry histories, company histories, industry research studies, business journals, trade journals, company correspondence, academic journals, research reports and newsclippings from the mass media. As was mentioned, Appendix 1 provides insight into the operationalization process of the most crucial concepts and the data sources used in them. As can be seen from Appendix 1, it was possible to acquire the same information from several independent sources. These sources are discussed in the next section.

3.4 Validity and Reliability of the Empirical Research

The purpose of this section is to discuss the quality of the empirical research. This quality will be manifested through the validity and reliability of the empirical study. The case study design and the other designs have their limitations. Yin (1984) divides validity in three parts: construct, internal and external. According to Yin, definitions of construct validity refer to the establishment of correct operational measures for concepts being studied. The analysis of the operationalization process of this study can be divided into two parts: (1) the evaluation of the background of the present author and (2) the quality of data sources. Since the present study concerns a technological change in process technology, the engineering (power electronics and automation) background and the working experience (around 10 years in industrial companies both in marketing and R&D areas) of the present researcher give a good platform for technologically oriented research. Moreover, the working experience is from other fields than the flat glass industry, thus permitting an unbiased understanding of the industries, firms and their actions. The experience of natural sciences requires a consistent approach concerning the empirical material.

3.4.1 Use of Multiple Data Sources

Multiple data sources were used in this study. As mentioned earlier, the data in this study were mainly collected from documentation, i.e. industry histories, company histories, industry research studies, business journals, trade journals, company correspondence, academic journals, research reports and newsclippings from the mass media. Enclosed is a brief analysis of the most pertinent sources.

Interviews. As was mentioned earlier, the role of the interviews during the teaching-case writing phase was to guarantee that a right and truthful understanding of the flat glass industry was achieved (see more Uusitalo 1995a, b). Thanks to these interviews a few fundamental facts (the existence of two independent subindustries, the invention of float glass in the plate glass industry, Pilkington's licensing policy and the role of the safety glass industry in the diffusion of float glass) were grasped at the beginning of the case writing process. Further interviews (either personnel of by telephone) were carried out in Finland and the UK. Sir Antony Pilkington, Chairman of Pilkington; Prof. Barker, President of the Historical Congress and the author of three histories, and Prof. Pearson, a former Pilkington researcher, were interviewed in summer 1995. The licentiate thesis (Uusitalo 1995a) was mailed to some persons who have worked or worked in 1995 either in the flat glass or safety glass industries. These persons were interviewed by phone in August and September 1995. Mr. Archinaco, Vice President, Glass in PPG Industries, has also commented on the licentiate thesis (the letter August 21, 1995). Appendix 2 provides a summary of the interviews and the comments on the licentiate thesis.

Company and industry histories. Professor Barker's (London School of Economics and Political Science /Department of Economic History) three histories (Barker 1960, 1977a, 1994) of Pilkington and the flat glass industry give a solid documentation of the flat glass industry, Pilkington and the diffusion of float glass. Other important and reliable histories are Hamon (1988), Daviet (1989), PPG (1967, 1983), Spoerer et al. (1987) and Saint-Gobain (1965). Hast (1991) provides a brief history of Asahi Glass, Boussois-Souchon-Neuvesel, Nippon Sheet Glass, Pilkington, PPG Industries, St. Gobain and Trinova (former Libbey-Owens Ford Industries).

Books on the Flat Glass Technology and Ph. D. Dissertations on the Flat Glass Industry. Four books (Doyle 1979, Persson 1969, Pincus 1983 and Tooley 1984) are cited in this area. Pat Doyle worked with the British Glass Industry Research Association in 1960–1979, and he has written several reports and literature reviews and contributed various articles to the technical press on subjects relating to glass technology and research. Rune Persson has worked since 1964 as R&D director within Grängesberg Co. (Gränges) in its glass division (Oxelösund and Scanglas) (Berg 1984:22). Alexis Pincus, a Technical Editor of The Glass Industry trade journal, has edited TGI articles concerning glass manufacture over a 60-year period. Prof. Tooley, Consultant, Professor Emeritus of the University of Illinois, has edited the glass technology articles in his book. According to the information given above, these sources can be regarded as reliable. In their Ph. D. dissertations Bain (1964) studied the impact of technological change on the flat glass industry, Frederiksen (1974) the effects of competition from abroad on the flat glass industry and Skeddle (1977) the investment decision making of both plate glass and float glass production lines.

Trade Journals. The case concerning the U.S. market is mainly based on an extensive review of The Glass Industry and Ceramic Industry Magazine trade journal issues for 1961–1984. The Glass Industry and Ceramic Industry Magazine trade journals are solid magazines, which have been published since 1920 and 1923, respectively. Other trade journals which have been consulted are The American Glass Review, European Glass Review (later Glass Review) and Chemistry and Industry. The information given in these magazines has been compared and the critical events have been checked and verified in at least two of them. The only differences noticed so far have been the time difference (2–6 months) of publication of an event, for example the introduction of PPG's float glass technology.

Business Magazines and Business Books. Several articles from business magazines (The Economist, Fortune, International Management and Management Today) provided the view of management. Since all these sources have existed for a long time they are regarded as reliable in this study. Two books (Berg 1984 and Grundy 1990) written by managers are worth mentioning. Bo Berg, manager of a Swedish and Swedish—Danish glass manufacturers, has written a rich story of the events in Scandinavia in the 1960s and 1970s. The information from Berg's book has been verified with other sources (statistics, newsclippings, etc.). Tom Grundy, a former employee of Pilkington, worked for the company 46 years, starting in 1933. Tom Grundy rose rapidly from the shopfloor to became a foreman and then in 1963 a manager at the Cowley Hill Works at Pilkington. According to Sir Alastair Pilkington, Tom Grundy has written the story of float glass right from the early development struggles, through to the first success and then to the spread of float glass throughout the world as the universal process for making flat glass. Sir Alastair Pilkington also writes that Tom(my) Grundy himself played a leading part for many years both during the development period and while we were teaching the world how to make float glass (Grundy 1990: Foreword).

Archival Records. Archival records (i.e. industry statistics, production volumes, import and export) were also used in the United States (Frederiksen 1974) and the Scandinavian market (Uusitalo 1997). The statistics can be regarded as reliable source. The documentation of the data and archival records are filed in chronological order. Appendix 1 shows how these data sources were used during the operationalization and research processes.

3.4.2 Analysis of Data

The analysis of the data is important in the case of explanatory and causal studies. The concept of internal validity deals with establishing a causal relationship, whereby certain conditions are shown to lead to other conditions, as distinguished from false relationships. Internal validity can be enhanced by doing pattern-matching analysis These sub-analyses were then combined in a way which resembled pattern recognition (Minztberg 1979), seeing evidence through multiple lenses (Eisenhardt 1989), and the use of data triangulation (Jick 1979) to construct case studies from a variety of information sources. This study can also be regarded as explanatory, as was mentioned earlier. The high level of internal validity in this study has been assured with subanalyses which approached the research phenomenon (i.e. the evolution of the flat glass industry during the research period) from many different perspectives. These subanalyses are combined the end of the analysis in a way which resembles pattern matching and explanation building.

Table 3.1 Viewing the research phenomenon from different perspectives

Perspective	Focus of analysis
Technology	Plate, sheet and float glass manufacturing processes
International technology transfer	Wholly-owned subsidiary, licensing or joint venture
Industry	The plate glass and sheet glass industries
Economy	
Large	The United States
Small	Scandinavia
Regional	
Concentrated	Plate glass industry in the United States and Europe
Fragmented	Sheet glass industry in the United States and Europe
Global	Licensing of float glass
Company	
Large (MNC)	Pilkington, St. Gobain, PPG, LOF
Small	Sheet glass manufacturers
Vertical integration	The safety glass
(value added industries)	
Network	Network of plate glass manufacturers
	Network of technologies

Table 3.1 provides a list of different perspectives. The basis for these perspectives was taken from Pettigrew and Whipp (1991) (see also Fig. 3.2) and McGee and Thomas (1988). However, some of perspectives emerged as a result of the interaction between empirical and theoretical inquiries.

As far as reliability is concerned, a critical point in a case study is that the operations of the study (data collection and analysis) can be repeated with the same results. According to Yin (1984:36), this is achieved by developing the case study protocol and case study data base. The objective is to be sure that later researchers, following exactly the same procedures as described by an earlier investigator and conducting the same case study all over again, will achieve the same findings and conclusions. (Yin 1984:34–36). The reliability of the empirical study is assumed to be high because the data were collected from many historical independent documents. Furthermore, the study protocol (i.e. first grasping the nature of the industry and then doing research) in a sense repeats the data collection since the new markets (The United States and Europe) were added with new empirical material from different, independent sources. Since the data consist of historical material, the research can be repeated quite easily and with the same results.

3.5 A Guide for Reading the Empirical Case Illustration

Since the purpose of the study is to evaluate whether the number of digits in SIC codes matters in the theory development, it is useful to provide a guide for reading the empirical case illustrations. These illustrations emphasize certain events and aspects which are derived from Fig. 1.6 and Table 3.1. The empirical cases in the next chapter, Chap. 4, stress the following:

- There were two branches, plate glass (SIC code 32111) and sheet glass (32112), in the flat glass industry as late as the mid-1970s.
- The state of the technologies in each branch before the invention of float glass.
- Float glass was invented in the plate glass branch and the UK based Pilkington managed with its licensing policy to prevent newcomers from entering the capital-intensive plate glass industry.
- Float glass was not introduced in the sheet glass branch until 1968-1970 because it was too expensive and too thick. Thus, the replacement of sheet glass by float glass was not a sudden event.
- The U.S. sheet glass market was extremely price-sensitive in 1960–1975, due to increasing imports of cheap sheet glass.

References

Anderson P (1988) On the nature of technological progress and industrial dynamics. Unpublished Ph.D. dissertation, Columbia University, New York

Ashton RL (1969) Solid state drive system at Ford's float glass plant. The Glass Industry, June, pp 303–306

Bain T (1964) The impact of technological change on the flat glass industry and union's reaction to change: colonial period to the present. Unpublished Ph.D. dissertation, University of California, Berkeley

Barker TC (1960) Pilkington Brothers and the glass industry. George Allen and Unwin Ltd., London

Barker TC (1977a) The glassmakers. Pilkington: the rise of an international company 1826–1976. Weidenfeld and Nicholson, London

Barker TC (1994) An age of glass. Pilkington: the illustrated history. Boxtree, London

Berg B (1984) Det Stora Glaskriget (The glass war). Glasmästeribranschens sevice AB

Cipolla CM (1991) Between history and economics—an introduction to economic history. Basil Blackwell, Oxford

Daviet J-P (1989) Une multinational a la Française. Histoire de Saint-Gobain 1665–1989. Fayard, Paris

Derclaye M (1982) La diffusion d'une innovation technique, le "float-glass" et ses consequences sur le marche du verre plat. Annales de Sciences Economiques Appliquees 38(1):133–148

Doyle PJ (ed) (1979) Glass making today. Portcullis Press Ltd, Redhill

Dubois A, Gadde L-E (2002) Systematic combining: an abductive approach to case research. J Bus Res 55(7):553–560

Earle KJB (1967) The development of the float glass process and the future of the glass industry. Chemistry and Industry, 15 July, pp 1197–1201

Edge KC (1984) Section 11, flat glass manufacturing processes (update). In: Tooley F (ed) Handbook of glass manufacture, 3rd edn, vols I and II, Books for the glass industry division. Ashlee Publishing Co., New York, pp 714/1–21

Eisenhardt KM (1989) Building theories from case study research. Acad Manag Rev 14:532–550

Frederiksen PC (1974) Prospects of competition from abroad in major manufacturing oligopolies: case studies of flat glass, primary aluminum, typewriters, and wheel tractors. Unpublished Ph.D. dissertation, Washington State University

Friedrich H (1975) Der technische Fortschritt in der Glaserzeugung. Eine Untersuchung über die Auswirkung des technischen Fortchritt auf den Strukturwandel in der Flachglasindustrie (Bochumer Wirtschaftswissenschaftliche Studien Nr. 7), Bochum

Grundy T (1990) The global miracle of float glass: a tribute to St. Helens and its glass workers. St. Helens, Merseyside

Hagg I, Hedlund G (1978) Case studies in social science research. The European Institute for Advanced Studies in Management. Brussels (Working Paper No. 78-16)

Hamon M (1988) Du Soleil a la Terre, Une Histoire de Saint-Gobain. Jean-Claude Lattès, Paris

Hast A (ed) (1991) International directory of company histories, vol III. St. James Press, London

Havard AD (1976) Flat glass forming. Glasteknisk Tidskrift 31(3):61–63

Jick TD (1979) Mixing qualitative and quantitative methods: triangulation in action. Adm Sci Q 24:602–611

Kinkead G (1983) The end of ease at Pilkington's. Fortune, 21 March, pp 90–92, 94, 96

Lilja K (1983) Types of case study designs. In: Proceedings of seminar on methodology in management and business research, Espoo, Finland

Lowry AT (1982) Pilkington: reflections on an uncertain future. Multinational Bus 3:18–30

Lundgren A (1991) Technological innovation and industrial evolution—the emergence of industrial networks. Dissertation, Stockholm School of Economics, Stockholm

McGee J, Thomas H (1988) Making sense of complex industries. In: Hood J, Vahlne J-E (eds) Strategies in global competition. Groom Helm, London, pp 40–79

Miles MB (1979) Qualitative data as an attractive nuisance: the problem of analysis. Adm Sci Q 24:590–601

Minztberg H (1979) An emerging strategy of "direct" research. Adm Sci Q 24:683–689

Patton QM (1990) Qualitative evaluation and research methods. Sage, Newbury Park

Persson R (1969) Flat glass technology. Butterworths, London

Pettigrew AM (1985) The awakening giant: continuity and change in ICI. Basil Blackwell, Oxford

Pettigrew AM (1990) Longitudinal field research on change: theory and practice. Organ Sci 1(3):267–292

Pettigrew AM, Whipp R (1991) Managing change for competitive success. Blackwell Publishers, Oxford

Pilkington A (1963) The development of float glass. The Glass Industry, February, pp 80–81, 100–102

Pilkington A (1969a) The float glass process. Proc R Soc London A314:1–25

Pilkington A (1969b) Glass and windows chance memorial lecture. Chemistry and Industry, 8 February, pp 156–162

Pilkington A (1971) Float: an application of science, analysis, and judgement. Turner Memorial Lecture. Glass Technol, August

Pilkington A (1976) Flat glass—evolution and revolution over 60 years. Glass Technol 17:182–193

Pincus AG (ed) (1983) Forming in the glass industry in two parts. Part one—forming machines and methods. Ashlee Publishing Co., New York

Porter M (1980) Competitive strategy. Free Press, New York

Porter M (1981) Strategic interaction: some lessons from industry histories for theory and anti-trust policy. In: Salop SC (ed) Strategy, predation and anti-trust analysis. Federal Trade Commission, Washington DC, pp 449–506

PPG (1967) Romance of glass. Public relations department, Pittsburgh

PPG (1983) A century of achievement. PPG Prod Mag 91(2)2:1–33

Quinn JB (1977) Pilkington Brothers LTD. The Amos Tuck School of Business Administration, Dartmouth College

Quinn JB (1980) Strategies for change. Logical incrementalism. Dow-Jones Irwin, Homewood

Saint-Gobain (1965) Compagnie de Saint-Gobain, 1665–1965: Livre du tricentenaire, Paris

Siggelkow N (2007) Persuasion with case studies. Acad Manag J 50(1):20–24

Simpson HE (1961) The glass industry—1960 a review. The Glass Industry, January, 67–70

Simpson HE (1963) The glass industry—1962 a review. The Glass Industry, January, pp 71–74

Skeddle RW (1977) Empirical perspective on major capital decisions. Unpublished Ph.D. Dissertation, Case Western Reserve University

Skriba DA (1971) How a computer and glass communicate. The Glass Industry, December, pp 442–445

Spoerer M, Busi A, Krewinkel HW (1987) 500 Jahre Flachglas, 1487–1987 Von der Waldhütte zum Konzern, (in German, 500 year old flachglas, 1487–1987, from a small business to a concern). Karl Hofmann Verlag, Schorndorf

Takahashi S, Ichinose M (1980) New vertical draw process for sheet glass. The Glass Industry, April, pp 24, 29–30, 32

Tooley F (ed) (1984) Handbook of glass manufacture, 3rd edn., vols I and II, Books for the glass industry division. Ashlee Publishing Co., New York

Uusitalo O (1993) Pilkington goes north: competition in Scandinavian flat glass market. An unpublished teaching case

Uusitalo O (1995a) The flat industry—the effects of the float glass on the industry structure. Licentiate thesis B-156, Helsinki School of Economics, Helsinki

Uusitalo O (1995b) A revolutionary dominant design - the float glass innovation in the flat glass industry. Dissertation A-108, Helsinki School of Economics, Helsinki

Uusitalo O (1997) Development of the flat glass industry in Scandinavia 1910–1990: the Impact of technological change. Scand Econ Hist Rev 3:276–295

Suárez FF, Utterback J (1995) Dominant designs and the survival of firms. Strateg Manag J 16:415–430

Vincent GL (1960) PPG operates largest, most modern U.S. plate glass plant. Ceramic Industry Magazine, September, pp 100–105, 136

Vincent GL (1962) How Ford produces plate in world's largest integrated glass plant. Ceramic Industry Magazine, November, pp 50–55, 80

Wierzynski GH (1968) The eccentric lords of float. Fortune, July, 90–92, 121–124

Yin RK (1984) Case study research. Design and methods. Sage Publications, Newbury Park

Zander U (1991) Exploiting A technological edge—voluntary and involuntary dissemination of technology. Dissertation, Stockholm School of Economics, Stockholm

The Glass Industry (TGI) trade journal: several articles and newsletters from the issues of 1960–1984

American Glass Review trade journal: Several articles and newsletters from the issues of 1971–1975

Ceramic Industry Magazine trade journal: several articles and newsletters from the issues of 1960–1980

Chemistry and Industry trade journal: several articles and newsletters from the issues of 1960–1975

Several newsclippings from international business magazines and newspapers (1960–1993)

Chapter 4
Flat Glass Industry in 1920–1980

Abstract This chapter illustrates the flat glass in industry in 1930–1990. Since the end of the 19th century the flat glass industry had two separate industries: the sheet glass and the plate glass sub-industries with different manufacturing technologies. Sheet glass was cheap and had optical distortions while plate glass was expensive having perfect optical quality. Sheet glass was used in ordinary construction and plate glass in mirrors, in architectural use and in cars as raw material for safety glass. The manufacturing technologies of both flat glass and safety glass are also described. This section culminates to float glass technology, a brilliant innovation introduced in 1959. Then the evolution of both sub industries in two markets, the United States and Europe, before the introduction of float glass is illustrated. Finally the diffusion of float glass on both industries and markets is described. Float glass entered first into the plate glass industry and after six to eight years into the sheet glass market converging two separate industries into one industry.

Keywords Convergence · Diffusion · Flat glass · Float glass

This chapter provides an extensive illustration of the evolution of the flat glass industry in the United States and Europe. The chapter contains five sections. First, there are brief descriptions of the main flat glass manufacturing technologies developed and used in the last century. In the first part of the twentieth century there were two different subindustries, plate glass and flat glass, in the flat glass industry. The former supplied high quality flat glass to be used as raw material for safety glass in the car industry and for architectural use in the construction industry. The industry was concentrated and required high capital investments. The latter one provided flat glass mainly for ordinary construction. Both large and small companies manufactured sheet glass. The second section provides a brief picture of one value-added product, safety glass, so pertinent to the flat glass producers. Next, the third section illustrates the evolution of both the plate glass and sheet glass subindustries in both the United States and Europe during 1920–1960. The fourth section describes the diffusion of float glass in both

O. Uusitalo, *Float Glass Innovation in the Flat Glass Industry*,
SpringerBriefs in Applied Sciences and Technology,
DOI: 10.1007/978-3-319-06829-9_4, © The Author(s) 2014

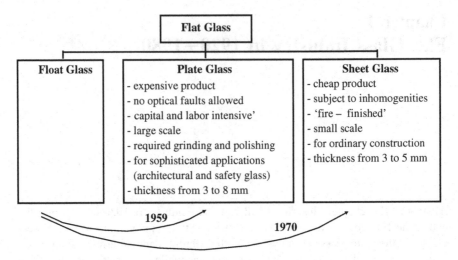

Fig. 4.1 Float glass entering the flat glass industry

subindustries within United States and Europe in 1960–1980. Brief summaries the main events of the diffusion of float glass in these markets are provided at the end of chapter.

4.1 Flat Glass Products and Technologies

This section examines briefly the flat glass products and technologies (including float glass) used since the 1920s in the two subindustries which existed as late as the early 1970s. These two quite separate subindustries in the flat glass industry were the sheet glass and the plate glass (Persson 1969; Doyle 1979, see Fig. 4.1). Sheet glass is divided in three categories (United States Tariff Commission 1968:2). The respective Standard Industrial Classification (SIC) and Customs Co-operation Council Nomenclature (CCCN) codes are as follows:

Type	SIC	CCCN
Flat glass	3211	70.05
Sheet glass	32111	70.05.100 (<4.0 mm)
		70.05.200 (>4.0 mm)
Plate glass	32112	
Float glass	32112	70.05.900

Sheet glass was cheap, 'fire-finished' (i.e. it did not require any processing after cooling) glass subject to inhomogeneities and optical distortion. Sheet glass was suitable for ordinary windows used in construction. Plate glass was needed in more

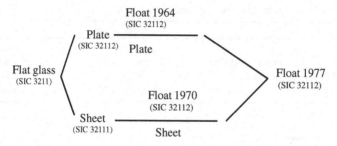

Fig. 4.2 The substitution of plate glass and sheet glass for float glass

sophisticated applications such as mirrors, automobile windows and the large windows used for retail displays and architectural effects, where the inhomogeneities and optical distortion were not acceptable.

Float glass was invented in the plate glass industry in the 1950s (see Fig. 4.1) by Pilkington Brothers Ltd. (later on referred to as Pilkington) to replace the labor- and capital-intensive plate glass process (Quinn 1977, 1979; Caulkin 1987; Arbose 1986; Salmans 1980; Wierzynski 1968). Figure 4.2 illustrates the U.S. substitution processes of plate and sheet glasses for float glass (see also Appendix 3). Figure 4.2 also illustrates the evolution of SIC codes during the convergence of two industries. Float glass was introduced in 1964 in the U.S. plate glass industry and it overtook plate glass by 1974. Float glass was introduced in the U.S. sheet glass industry in 1970 and by 1977 it had replaced almost all existing sheet glass processes. That time also the Bureau of Census stopped distinguishing the plate glass and sheet glass industry by the SIC codes.

Since 1984 all flat glass made in the United States and Europe has been produced with the float glass process. By the turn of the twentieth century, sheet glass was manufactured with the cylinder process. This process involved blowing a large cylinder which was allowed to cool before being split and flattened (Pilkington 1969a:2). The blowing phase was mechanized by American Window Glass company in 1903. The machines were called Lubbers according to the inventor. However, machine blowing did not entirely replace hand-blowing (Anderson 1988; Barker 1977a).

Continuous sheet glass production methods were invented independently in the early 1900s both in Europe and the United States (Doyle 1979:187) as shown in Table 4.1. Sheet glass was drawn into a ribbon through a block floating on the surface of the molten glass inside a glass furnace. The ribbon passed vertically upward through an asbestos roller, a 'lehr' which relieved the stresses in glass and then into a cutting room where the cooled, hardened glass was cut and stacked (see Fig. 4.3). The Fourcault process was good for making thin glass (1.5–7 mm). In the Colburn process a glass ribbon is drawn vertically from molten glass and bent before cooling at a height of one meter above horizontal rollers. The rights to the Colburn process were later acquired by Libbey–Owens (the predecessor of Libbey–Owens–Ford Glass and Libbey–Owens–Ford Industries, hereinafter referred to as LOF).

Table 4.1 Continuous sheet glass (vertical draw) processes

Process	Introduction	Origin
Fourcault	the early 1900s	Belgium
Colburn or Libbey–Owens–Ford (LOF)	1906	United States
Pittsburgh (PPG) or Pennvernon	1926	United States
Asahi	1971	Japan

Fig. 4.3 The drawing process

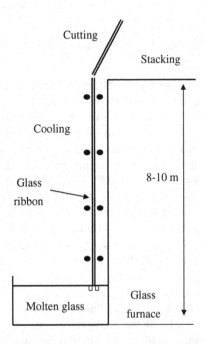

Consequently the Colburn method was also called the Libbey–Owens method. This process had approximately the same characteristics as the Fourcault process. In 1926 Pittsburgh Plate Glass, later on PPG Industries (PPG), introduced the Pittsburgh (also called Pennvernon) process which combined the advantages of both the Fourcault and the Colburn methods. The Pittsburgh process was similar to the Fourcault process, but it was more effective than the Fourcault for the production of thick glass. In 1971 Asahi Glass Co. (Asahi) introduced its vertical drawing process, which was an improvement of the Fourcault method (Uusitalo 1995b:76; Takahashi and Inchinose 1980).

To make plate, molten glass was rolled into a plate with a waffled surface and then ground and polished until both surfaces were smooth and parallel. In 1923 Pilkington together with Ford Motor Co. (Ford) installed the industry's first continuous plate manufacturing process at St. Helens. 12 years later Pilkington introduced a 'twin' machine to grind both sides of a plate glass ribbon simultaneously. Twin grinding gave Pilkington a technological advantage in plate glass manufacturing (Barker 1994:79). The twin grinding technology was first licensed

in 1937 to the leading continental manufacturer, the St. Gobain Co., for non-
exclusive use in France and its territories, and for exclusive use in Germany, Italy
and Spain where St. Gobain also had factories (Barker 1977b:189). Later on the
twin grinding technology was licensed to Glaver (the predecessor of Glaverbel)
from Belgium, Boussois from France, LOF, PPG and ASG from the United States
in 1950, 1954, 1951, 1954 and 1960, respectively (Barker 1994:79; Barbour
1971:7). Thus, twin grinding was a technological discontinuity in the United States
in 1951 when LOF took it in use. In 1962 while opening its plate glass plant, ASG
introduced in the United States St. Gobain, a twin polishing method called Jusant
developed by the parent company St. Gobain in France (Hamon 1988:142;
Simpson 1963:73; Cover Story 1962:597–601).

The difficulties and costs in the production of plate glass were well known in
the industry as the following quotation shows. "Although the twin grinder was
hailed as a great technical feat, it was a feat of engineering rather than of glass
making. The whole plant - tank furnace, annealing lehr, twin grinder and polishers
- stretched out in a line no less than 1,400 feet (420 m) long. Contemporaries noted
(whether with pride or regret is not clear) that this was seventy feet longer than the
Queen Mary, then the largest ship afloat. The fixed capital costs of electric motors
and machinery were enormous. The running costs were also considerable. A
supply of electricity amounting to 1,500 kilowatts was needed to grind 10 % off
each side of the plate glass ribbon as it emerged from the lehr, not to mention the
cost of the extra fuel and raw materials needed in the first place" (Barker 1994:79).

The float glass process was introduced by Pilkington in 1959. In the float glass
process, a continuous ribbon of glass moves out of the melting furnace and floats
along the surface of an enclosed bath of molten tin (see Fig. 4.4). The ribbon is
held in a chemically controlled atmosphere at a high enough temperature for a long
enough time for (the irregularities to melt out and) the surfaces to become flat and
parallel. Because the surface of the molten tin is dead flat, the glass also becomes
flat. As a result of this process it became possible to make thinner flat glass of high
quality while eliminating the grinding and polishing steps required in other
manufacturing methods. The process was estimated to be ten times more efficient
than the old grinding and polishing technique (Phoenix Award Winner 1981:16).

Pilkington (1963) (See Appendix 4) provides some of the features of the float
glass manufacturing process at the time it was introduced in the U.S. market. Float
glass advanced the plate glass industry a huge step forward towards a real process
industry. The length of the production runs was increased substantially. The
increase in the productivity of a float glass line was due to the elimination of a
huge number of rotating components needed in the grinding and polishing units of
a plate glass line. The customer and warehousing requirements could be taken into
account more effectively since the line was easier to control. The timing of the
improvements in process control techniques (thanks to the semiconductor technology) and control equipment could not have been better.

In 1975 the U.S.-based PPG introduced (Uusitalo 1995a:95) its own float glass
process (not violating the patents or rights of Pilkington's float glass process,
McCauley 1980; Perry 1984). In the present study the term float glass means only

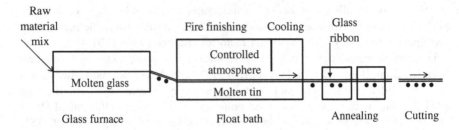

Fig. 4.4 The float glass manufacturing process

the float glass made with the process invented by Pilkington. When PPG's own float glass is discussed it is designated floatPPG glass. The next section provides an illustration of one value added product, safety glass.

4.2 The Processed Products, Safety Glass

Processed products are further processed from flat glass that is some value has added on flat glass. In the present study one important value added product, safety glass, has been taken into account. Safety glass is further divided into two types, depending on the manufacturing process: tempered and laminated. Safety glasses are basically divided into toughened glass and laminated glass. Tempered glass is most often used for safety glass because it is less expensive, while laminated glass is used mainly for more demanding conditions, i.e. for security and for the automobile industry in windshields. Toughened glass is suitable for the auto industry in side and rear windows. The rigid automaker specifications and motor vehicle safety standards as to glass, impact strength, optical quality, breakage characteristics and durability, etc., have to be met. The first trademarks in Europe for tempered and for laminated windscreen were Sécurit, Toughened and Triplex, respectively. The respective Customs Co-operation Council Nomenclature codes (CCCN) are as follows:

Type	SIC	CCCN	Use
Safety gass	31113	70.08	
Tempered		70.08.010	Vehicles
"		70.08.030	Bent, other use than vehicles
"		70.08.100	Other
Laminated		70.08.020	Vehicles
"		70.08.200	Other use than vehicles

Thermal Strengthened Glass. It has long been known that the strength of glass can be increased by heating it to a high temperature and then cooling it rapidly. In the 1920s St. Gobain introduced Sécurit, the trademark for tempered safety glass.

Glass must be heated to a temperature above its annealing (annealing means controlled cooling of glass to prevent stresses) point, usually 650 °C. It is then cooled rapidly with an air jet. The surface of the glass will then be in compression and balanced by tension from inside. Thus, permanent stress is set up in glass.

Thermally-tempered glass (toughened glass) cannot be broken unless sufficient force is applied. This force will first overcome the compression at the surface and introduce enough tension for breakage. A crack will always start at the surface and since the surface in tempered glass is in tension, its strength will be much higher than that of normal (untempered) glass. Tempered glass is usually three to five times stronger than normal glass with respect to impact, sustained load and thermal shocks. When thermally tempered glass breaks it fractures into a great many small pieces. These small pieces are interlocking and have no large jagged edges. They are therefore not likely to inflict serious injury. Tempered glass cannot be cut. Exact sizes must therefore be specified to tempered glass manufacturers (Persson 1969:96–98).

Laminated Glass. In general, laminated glass is made of two or more panes of glass, firmly united to each other by an interlaying reinforcing material, usually plastic (i.e. polyvinyl butyrate). Laminated glass was discovered by a French chemist, Edouard Bénedictus. He glued a celluloid film between two glass panes. In 1936 the celluloid, which over time became yellow, was replaced with a plastic film. The strength in bending or impact is usually somewhat less than that for a solid glass of equal total thickness. When laminated glass fractures the glass fragments are stacked into the interlaying plastic material. The lamination process is carried out in an autoclave at a pressure of 2–6 Kg/cm^2 and a temperature of approximately 100 °C. Laminated glass is used for automotive windshields and many architectural purposes (Persson 1969:98–100).

4.3 The Flat Glass Industry in 1920–1960

This section seeks to provide a concise picture of the flat glass industries in 1920–1960. The main differences between the products of the two branches, plate glass and sheet glass, were the optical quality and the price. Plate glass was distortion free and expensive while sheet glass was subject to distortions and inhomogeneities, but it was much cheaper than plate glass.

4.3.1 The Plate Glass Industry in 1920–1960

This section includes two parts, first, the evolution of the plate glass industry in 1920–1960 is illustrated and second, the origins of the safety glass industry are described. The safety glass industry is included because it had such important effects on the development of the plate glass industry.

The Plate Glass Industry. The plate glass industry was characterized by high quality and high price products and large capital-intensive investments. The Belgium manufacturers intensified their attempts to regulate international sales of plate glass as early as 1904, when the Convention Internationales des Glaceries (later on Convention) was formed in an effort to control the Continental market, by then over-supplied and unprofitable (Barker 1986:184). The Convention's policy was to discourage manufacture elsewhere for as long as possible and to regulate markets among themselves so as to run their machines at high and efficient levels. Pilkington had been outside the Convention and it had established a manufacturing unit at Maubeuge in France in the early 1920s. This unit sold 90 % of its production in France. In 1929 Pilkington joined the Convention which, in 1934, also reached an agreement with the U.S. plate glass manufacturers, PPG and LOF, to allocate world markets. According to the agreement, the markets outside the United States were divided so that the U.S. producers could take 20 % of them and the European producers the remainder (Barker 1977a:361). In 1935 Pilkington sold Maubeuge to St. Gobain, a large French producer.

As was mentioned earlier, Ford Motor Co. and Pilkington cooperated in inventing the first continuous casting (called the flow process) and the first continuous grinding line in 1923. Molten glass was taken continuously from a tank instead of being pouring in from a pot. With this new process, it was possible to produce thinner rough glass (8.6 mm instead of 10.6 mm with the old method). This meant extensive savings in energy on grinding when the required thickness of 6.4 mm (or 1/4″) was made (Barker 1977a:277). By 1927, the ribbon width was increased to 100″. By the end of 1924, Pilkington terminated the co-operation agreement with Ford to keep secret its new inventions in continuous plate glass manufacturing. Meanwhile, PPG in the United States also invented a similar continuous flow (from a tank) and a grinding process in 1923.

In Europe since 1922, Ateliers Heuze Malevez et Simon (HMS), the predecessor of Boussois (Grundy 1990:83), worked with continuous grinding and polishing technology. In 1923 Pilkington and the French HMS pooled their resources. Pilkington could exploit the Heuze–Pilkington technology (continuous grinding and polishing) exclusively in the UK and also in Maubeuge, France. HMS was responsible for the manufacturing and licensing of the machines on the Continent to the Convention (Barker 1977a:279). The Convention reached agreement in 1924 (Barker 1977a:279) and it installed this continuous grinding and polishing technology at Auvelis (St. Gobain), Chantereine (St. Gobain) and Boussois (Glace de Boussois) in France, at Herzogenrath (St. Gobain) in Germany, and Franiere (St. Gobain/St. Roch) and Moustier (Glaverbel) in Belgium.

The National Plate Glass Company and Libbey–Owens–Ford (LOF), after being formed in a merger in 1930 of Libbey–Owens and Edward Ford Glass Company of the United States, also licensed the Heuze–Pilkington technology in 1927 and 1930. It was not the continuous grinding but the continuous flow from a tank that gave the competitive advantage to Pilkington compared with its Continental rivals. Co-operation with HMS allowed Pilkington to keep the flow process secret (Barker 1977a:280).

In 1932 due to declining sales in the United States, PPG increased its presence in Europe. The Courcelles Works (acquired in 1902) in Belgium was useful for this purpose (AGR, November 1973:11). In 1937 St. Gobain had acquired a non-exclusive twin grinding license from Pilkington to France, and French territories abroad, and an exclusive license for Germany, Italy and Spain (Barker 1994:79). Before 1950 a few foreign direct investments (FDI) were made in the plate glass industry. St. Gobain had operated in Germany (West Germany) since 1855 (Daviet 1989:112) with the exception of the years during the World Wars. In the mid-1950s St. Gobain acquired the controlling interest in Glacerie St. Roch, a Belgium manufacturer. The high costs of production facilities prevented FDIs in the plate glass industry. However, at the end of the 1950s, St. Gobain decided to invest in a plate glass plant in the United States.

Since 1920s the expanding and internationalizing auto industry has been the main customer for the plate glass industry, because most safety glass is made from plate glass. Plate glass producers had already recognized this important industry in the 1920s. However, in the 1930s thick drawn sheet glass (TDS), which was also polished, started to compete with plate glass (Barker 1977a:370; Uusitalo and Möller, 1997). According to Barker (1977a:371), in the 1930s in all countries outside the Britain and the Commonwealth the plate glass industry, by controlling the safety glass companies, insisted on a continued supply of laminated safety glass for windshields and using thin plate glass for the purpose. To guarantee plate glass sales in the 1930s, Pilkington had to internalize the safety glass manufacturing in the UK, Canada, Australia and South Africa (Barker 1977a:379–388). Because of the strategic importance of the internationalizing (or globalizing) auto industry, the plate glass industry also faced a great deal of pressure to globalize.

Many companies invested in the plate glass process in the late 1950s. BSN and PPG started new lines in 1956 and 1957, respectively (Friedrich 1975:49 and Skeddle 1980:557). With the high capital costs which the introduction of continuous machinery entailed, new entry into this industry became both difficult and unattractive. World demand for plate glass could be satisfied by a handful of large producers located in the main industrial countries. Their intention was to use this costly machinery as intensively as possible, primarily to supply home demand but also to export as much as possible. (Barker 1986:184).

The plate glass industry was much more concentrated than the sheet glass branch. By 1960 four large companies, Pilkington from the UK, Boussois and St. Gobain from France and Glaverbel from Belgium were producing plate glass in Europe. In the United States there were three producers: Pittsburgh Plate Glass (PPG), Libbey–Owens–Ford Glass (LOF) and Ford Motor Co. (Ford).

The Events in Safety Glass. In the 1920s, when both the auto industry and the amount of glass per car grew rapidly, the demand for safety glass also increased rapidly. Safety glass was used to prevent from serious injuries in car accidents. The first patent for laminated safety glass was filed in 1909 by a French Eduard Benedictus (Barker 1977a:346–347). In 1912 Triplex was established in the UK. Triplex used as raw material both Pilkington plate glass and imported thin sheet glass from Bohemia and Belgium. In 1915 Triplex again offered unissued shares to Pilkington without any result.

In the mid-1920s Daimler, Riley, Lagonda, Rover, Fiat, Austin, Lanchester and Wolseley had all agreed to fit Triplex safety glass as an optional extra. At this moment Triplex started to compete with Pilkington glass in an important growing market. Pilkington had at the same time started continuous manufacturing of distortion-free plate glass. For Pilkington it was of the utmost importance to get the local safety glass producer to use its plate glass instead of cheap sheet glass. At the end of 1920s Triplex grew rapidly. In 1930 more safety glass was made of cheaper sheet glass than of the more expensive plate glass upon which Pilkington depended. During that time there were 30 safety glass producers in the UK. Since Pilkington had no sheet glass process, the sheet glass was imported from Belgium or Czechoslovakia. This position changed when Pilkington acquired the PPG process. Since 1937 it has been compulsory to have safety glass in car windows.

In the 1930s, a new sort of safety glass, tempered, was also introduced. This type of safety glass did not required such thin plate glass as a raw material as laminated safety glass required. Plate glass was much more competitive with sheet glass in tempered safety glass than in laminated safety glass. Tempered safety glass was invented in France by Boussois and St. Gobain. In 1930 Pilkington acquired the rights in the UK. (Barker 1977a:353). In the mid-1930s all Triplex factories bought Pilkington glass as long as Pilkington could match foreign competitors on price and delivery. In 1934 more tempered than laminated safety glass was sold in the UK. Increasing motor industry demand for plate glass was of paramount importance to Pilkington in the 1930s (Barker 1977a:354).

The production of safety glass became an important new branch of the industry (Barker 1994:346). In the United States Ford Motor Co. had integrated both into the flat and safety glass production. PPG (Trademarks Duplate for laminated and Herculite for tempered safety glass) and LOF started to produce safety glass in 1930 and 1931. LOF had acquired the American Triplex. The two U.S. plate glass giants saw that only plate glass was used for safety glass manufacturing in America (Barker 1977a:349). Quarter inch tempered safety glass was used in car side and back windows, which increased the demand for plate glass of this thickness.

4.3.2 The Sheet Glass Industry in 1920–1960

This section briefly illustrates the sheet glass industry in 1920–1960. As was mentioned in the technology section, two sheet glass drawing processes, Fourcault and Colburn, were invented independently in Europe and the United States in the early 1900s. After the First World War, these two processes and the Pennvernon process (announced in 1926 by PPG) started to replace the hand and machine blowing window glass processes both in Europe and the United States. By 1930, all sheet glass in the United States was produced by these drawing processes. In Europe some 5 % was still produced by the old cylinder methods.

In Europe, the drawn sheet flat processes diffused rapidly especially in Germany and Belgium. Mechaniver, a Belgian company, was granted exclusive rights for the Colburn (of LOF) process in Europe. The Fourcault process was, on the contrary, licensed liberally. Germany soon became a leading sheet glass producer in Europe and Belgium took second place. The increase in the number of Fourcault machines in Detag (in Germany) demonstrates this growth. In 1932, Detag had 31 Fourcault machines, while 12 years later the number of machines was 67 (Spoerer et al. 1987:146). The U.S. based technologies, Colburn and PPG, were not licensed in the United States to other sheet glass firms (Anderson 1988:342–343). Fourcault, on the contrary, was also widely licensed in the U.S. flat glass market.

In the late 1920s, Pilkington had major problems with sheet glass production. By 1910, it had acquired a license from American Window Glass (AWG) for machine blowing. Pilkington did not manage to license either the Fourcault or Colburn technology. In the early 1930s, when the Fourcault and Colburn process were working well, Pilkington still produced sheet glass with the blowing methods. However, at the last moment, Pilkington managed to acquire a license for the Pennvernon process. During the Second World War, exports of sheet glass to the UK declined, and the dominance of the Belgian producers came to an end. By 1930 there were 520 Fourcault machines, 37 LOF machines and a few Pennvernon machines in Europe (Friedrich 1975:50). In the late 1930s, Pilkington gained a reputation for being the most efficient operator of PPG machines in Europe. It became also the largest user of PPG machinery outside the United States (Barker 1977a:364).

After the war, capacities were increased. The Pennvernon technology began to supersede the Colburn and Fourcault processes. The production capacities of an individual sheet glass tank increased significantly. For instance, one of Pilkington's tanks in St. Helens, which in the early 1930s produced the 650,000 square feet per week with the first PPG machines, was making 1,150,000 square feet per week by 1954 and 1,800,000 square feet per week by the mid-1960s (Barker 1977a:413). The capacities of the Fourcault machines also increased. Detag announced in 1957 that the production of its 27 drawing machines was the same as that of 64 machines just after the war (Friedrich 1975:62).

Pilkington had produced sheet glass in South Africa, India and Canada since the early 1950s. St. Gobain had been in the U.S. sheet glass market since the mid 1920s. By early 1960 Germany was the largest producer of flat glass in Europe and Belgium was the second largest (Ohlin 1965). As was mentioned above, few different process technologies existed and all of them were licensed to other manufacturers with some exceptions in certain markets.

4.4 The Evolution of the Flat Glass Industry in 1960–1980 in Two Markets

This section, which illustrates the evolution of the flat glass industry in 1960–1980, has two sub-sections. In the subsections the evolution of the flat glass industry in 1960–1980 is reviewed in two different markets: The United States and Europe (see Fig. 1.6). Since I test whether the number (either four or five) of digits of SIC code in the definition of an industry matters I have disclosed the SIC codes of the industries in the analysis of the U.S. market.

4.4.1 The U.S. Flat Glass Industry (SIC 3211)

At the end of the 1950s three major manufacturers, PPG, LOF and Ford dominated the flat glass and safety glass market in the United States (see Table 4.2). Fourco Glass Co. is a single company in the marketplace which operates on the behalf of two independent sheet glass manufacturers, Rolland Glass and Harding Glass corporations (United States Tariff Commission 1967:23).

4.4.1.1 The Plate Glass Industry (SIC 32112)

In the 1950 three major manufacturers, PPG, LOF and Ford, produced all the polished plate glass in the United States All three were also vertically integrated downwards into safety glass production. For instance in 1960, LOF was the largest safety glass manufacturer in the world. Ford typically produced safety glass from its plate glass for use by Ford Motor Co.'s auto industry. In the 1960s, the rapidly growing auto industry via its safety glass purchases was obviously the main customer for the plate glass industry. In 1957 PPG opened a new, modern plate glass plant (Works #7) to produce industry prime product, 1/4″ mirror quality plate glass, in Cumberland (Vincent 1960:100).

In 1959 American Saint Gobain (ASG), a subsidiary of St. Gobain, France, began construction of a $40 million plate glass plant in Greenland, Tennessee (Hamon 1988:142). In 1961 Ford completed a multi-million expansion at its Nashville plate glass plant (Vincent 1962). In October 1962, when it opened its large plate glass plant (a final investment of $50 million and the capacity of 40 million square feet per year), ASG expected to acquire a share of some 10 % in the plate glass market in the United States. St. Gobain, the parent company, had developed a twin polishing method called Jusant; it claimed that the method would provide a much more economical method for polishing glass than the conventional system (Simpson 1963:73). The method was used at the new plant in the United States (Hamon 1988:142; Cover Story 1962:597–601). One person interviewed at the dedication said that the line was already old-fashioned (Daviet 1989).

Table 4.2 Major plate glass, sheet glass and safety glass producers in the United States at the beginning of the 1960s

Company	Plate glass (SIC 32112)	Sheet glass (SIC 32111)	Safety glass (SIC 32113)
PPG	×	×	×
LOF	×	×	×
Ford	×	×	×
American St. Gobain	×	×	×
Fourgo Glass		×	
Mississippi Glass		×	
Pilkington, Canada		×	
Permaglass			×
Guardian Ind.			×
Shatterproof			×

In July 1962, PPG was the first company in the world to license Pilkington's float glass process in the U.S. market. LOF followed the next year (Barker 1994). PPG's first float glass plant was erected in Cumberland, Maryland. LOF's new float glass line was built next to its automotive glass fabricating plant in Lathrop, California. PPG started to produce saleable float in early 1964 (Grundy 1990:58–59). At the opening ceremony in March 1964, Mr. Barker, the president of PPG, said that if float glass prices were reduced, it was very unlikely that comparable pieces for plate glass could maintain a higher price level, because "plate glass cannot be sold at a premium figure over float." At the same time Mr. Barker said that plate glass was already doomed. PPG had also announced its decision to build a second float glass line and the third float glass line was already under design (Anonym 1964:176–178).

The nominal or natural thickness of float glass is 6.4 mm (1/4″), which was also the thickness used most in the U.S. automobile industry at that time. Float glass therefore quickly acquired a dominant position as the raw material for the safety glass supplied to the automobile industry. (Quinn 1977 and discussion with Mr. Artama, 4 February 1993). In 1965 Ford converted its plate glass plant in Nashville, Tennessee, to the float glass process. In 1966 Pilkington started to build a float glass line in Canada. It was dedicated next year. In 1967 the Glass Industry trade journal (Uusitalo 1995b:95) described the applications of float glass as follows:

> Now being sold for glazing in applications where plate glass is normally used, float glass is appearing in schools, offices, buildings, and increasingly in residential construction - especially those in which thicker glass has to be used for picture windows. Float has been accepted by safety glass manufacturers throughout the world for production of both laminated and tempered glass for automobiles - windshields, side lites, and back lites.

In the early 1968 Pilkington introduced a method for tinting float glass within a normal production line. The company also stated that as little as one mile (i.e.

100 tons of glass ribbon, 1/4″ thick and 10 ft wide) of glass with special characteristics could be produced profitably. The tint of float glass could be changed by
a dial setting and the actual changeover took about five minutes. (Anonym
1968:20–21). In August 1970, a group of private investors from the United States
acquired St. Gobain's (France) holdings in ASG. Soon after the acquisition, ASG
Industries Inc. (ASG Ind.) wrote off two of its sheet glass plants and the grinding
and polishing units in the plate glass line and converted the plate glass line into a
float glass operation (Uusitalo 1995b:95). In October 1971, ASG Ind. started float
glass production in Greenland. 3 years later (in October 1974) the company shut
down its plate glass line.

In October 1970, Guardian Industries (Guardian) was the first company outside
the plate glass industry to introduce the float glass process (Kinkead 1982). The
process was obtained through a licensing agreement with Pilkington, which was
not signed until June 1971. Previously, Guardian had fabricated safety glass for
automotive use. In 1969 it merged with Permaglass, a US based safety glass
manufacturer and was listed in American Stock Exchange (Uusitalo 1995b:95).
The number of float glass lines with a Pilkington license grew rapidly in the 1960s
and the early 1970s (see Appendices 3 and 5). In 1975 PPG announced that it had
developed another type of float glass manufacturing process (Uusitalo 1995b:95).
The PPG process differs from that of Pilkington mainly in the furnace and forming
operations (McCauley 1980:20). It was said to produce more uniform thickness, as
well as higher optical quality in a full range of commercial thicknesses for the
transportation, construction and other flat glass markets.

4.4.1.2 The Sheet Glass Industry (SIC 32112)

In the early 1960s, all three major plate glass manufacturers, PPG, LOF and Ford,
were in the sheet glass business. The newcomer in plate glass, ASG, also produced
sheet glass in its three plants (Arnold, Jeannette and Okmulgee). Ford used all the
sheet glass it produced internally (Uusitalo 1995b:96). There were also some
smaller sheet glass manufacturers in the market (Fourco Glass Co., Mississippi
Glass Co.).

In 1960 the advances in manufacturing techniques in drawing molten glass had
made it possible to produce improved sheet glass that was considered superior
according, to the standard distortion measuring device, to any sheet glass previously manufactured in the United States. Known as 'Premium Pennvernon,' the
glass was free from distortion and was manufactured on the Pennvernon machine.
The product was expected to find a high degree of acceptance among architects,
sash manufacturers, automobile manufacturers, and manufacturers of flat glass
specialty products. Noticeably more attractive in a building installation than other
sheet glass, Premium Pennvernon was the most nearly optically perfect glass then
available for architectural uses—short of the ultimate perfection obtained only in
ground and polished plate glass (Simpson 1961:69).

In spring 1961 the Tariff Commission (TC) had recommended to the President Kennedy that the tariffs for sheet glass should be increased. The sheet glass industry had presented the following information to the TC. In 1950 domestic producers supplied 98 % of the demand. 10 years later imported sheet glass had taken a 25 % share of the market. Sale of imported glass at lower price had weakened the domestic price structure. In 1960 four companies accounting for the bulk of the domestic output of sheet glass had shown a net operating loss of $1.2 million while 5 years earlier the aggregate net operating profit in the industry had been $30 million.

According to the commission, the principal factors leading to the import trend were periodic shortages, availability of thinner and cheaper single- and double-strength glass from foreign manufacturers and the reluctance of U.S. producers to bypass their direct factory-distributor customers in order to sell directly to other large volume distributors and industrial users. The Commission had the following to say, particularly about distribution (The Tariff Situation 1961):

> The U.S. producers sell their glass to carefully selected distributors and jobbers, to fabricators such as sash and door manufacturers, and to processors such as temperers, laminators and mirror manufacturers. These so-called reorganized factory buyers, selected according to the judgment of the individual producers, are the only concerns that can buy sheet glass directly from factory. Other concerns desiring to purchase sheet glass, even in carload lots, must order their glass from the recognized factory distributors at correspondingly higher prices.
>
> Under this distribution system the glass may be sold down through succeeding business levels (and at correspondingly higher prices) beginning with the recognized factory buyer and followed by the smaller distributor or jobber, the dealer, and the retailer. The processor and fabricator, and the building and glazing contractors, depending on their size, buy at different levels. The direct factory buyer classification, however, is carefully controlled by the U.S. producers.
>
> As a result of this control, jobbers and distributors that are not recognized factory buyers (even though they may purchase in carload quantities) have to compete with the distributors and jobbers that are so recognized and from whom they are forced to buy their glass. This element of competition between the higher and lower levels of distribution of domestic glass has encouraged the importation of sheet glass. Those firms at the lower levels of distribution can import directly at the same price as the domestic factory buyers and can thus compete with the latter on equal terms.

In the hearing J. B. Booth of Seaply Glass Corp., New York and William J. Barnard, a Washington attorney, represented Finnish (and Austrian) interests. Booth concluded,

> The domestic industry can solve most of its difficulties by making what is marketable instead of marketing what they make.

In June 1962 higher tariffs for imported sheet glass were in effect. The increase corresponded to a more than 5 % increase in the selling price (Sheet glass and the tariff 1962). In March 1964 at the opening ceremony of the first float glass plant, Mr. Barker, the vice president of PPG, mentioned that float glass cannot compete today with sheet glass, since float glass is not so much better than sheet glass as to merit paying the price differential. To cite an example, PPG was taking its

Premium Pennvernon off the market because of the high quality possible in sheet glass. In other words, the sheet glass market was highly price-sensitive.

In his analysis of the U.S. sheet glass manufacturers, Mr. Barker, the vice president of PPG (Barker 1964), mentioned the following reasons for the current difficulties:

- the successive reductions in rates of duty applicable to sheet glass that took place over the years were a substantial factor
- the great disparity between wage rates paid to sheet glass workers among the producing countries in the world
- the cost advantage in raw materials (according to a cited study the costs in the Common Market were some 15 % lower than those in the United States) enjoyed by foreign producers
- the companies which imported to the United States were large, well performing organizations.

Sheet glass producers in the United States invested heavily in their plants. ASG's sheet plants went through a $3 million modernization program in 1962. In the mid-1960s, PPG erected totally new sheet glass plants or improved the existing facilities. The Mount Vernon plant's warehousing was modernized in 1964. At the Mount Zion, Illinois plant two drawing machines were added on two occasions (increasing the capacity first by 50 % and then by 30 %) in September 1965 and in January 1967. In the opening speech at Mount Zion, Mr. Barker, the vice president of PPG, said that further expansion plans would depend on action taken by the U.S. Tariff Commission. Fresno, California with six Pennvernon drawing machines (Allen 1967) was dedicated in 24 July 1967. The fifth domestic sheet glass plant for PPG was considered one of America's most modern facility and it 'is serving as a model for glass plants the company will build elsewhere' (Allen 1967:51). Owen Sound also with six Pennvernon (Allen 1968) was inaugurated in Ontario, Canada in February 1968. The Canadian investment was $20 million.

The tariff question was discussed constantly. In 1967 TGI (Editorial 1967) wrote: "it would appear that President Johnson's decision to reduce and partly eliminate the escape-clause duties on sheet glass is based on obsolete information, and not on the facts of today." Further the TGI said that the tariff should not have been reduced in the first place. The tariffs would have to be reinstated immediately if the imports were to increase. At the annual meeting of the Flat Glass Marketing Association in June 1967, Mr. Wingerter, president of LOF, assailed the Government's tariff policy. The window glass operation was 'in a struggle to stay alive.' LOF was not making money, and had not recorded a satisfactory performance in window glass for several years. But, in an apparent about-face, he assured jobbers that LOF was not considering going out of the window glass business. He continued "We simply aren't going to continue spinning our wheels unprofitably." LOF could import window glass from foreign companies in which it had a financial interest, it could build a new plant and abandon present operation or, "we can work like hell to revamp and improve the efficiency in our present plants," Mr. Wingerter said. (F.G.M.A. Convention Report 1967:63). The situation was also reposterd to the President of the United States (United States Tariff Commission 1968).

In the 1950s the import of sheet glass grew from 3 to 22 % of the U.S. sheet glass consumption. In the 1960s it levelled to 25 %. The maximum import, 32 %, was hit in 1968 (Frederiksen 1974). In 1968 the sheet glass branch was not regarded as inferior to the plate (or the float) glass branch. Morgan Worcester, a significant supplier of the glass (both flat glass and container glass) industry, referred in its full page advertisements in TGI to PPG's new window glass factory at Fresno, California. Glaverbel, a Belgium company, also heavily advertised in TGI, boasting that it had the largest drawn sheet glass tank in the world. In October 1969 U.S. sheet glass manufacturers again urged the Tariff Commission to prevent sharply rising glass imports from capturing a major share of the American market.

In July 1969, PPG announced that it had developed a new vertical manufacturing technology which enabled the company to produce thin flat glass (1/8″ and less in thickness) that would compete in quality and cost with float glass (Uusitalo 1995b:99). In December 1969, the company published a catalogue (12 pages) of the 'new' Pennvernon Sheet Glass. The pamphlet described the mechanical, thermal and optical properties of Pennvernon transparent and Graylite glasses (Uusitalo 1995b:99). In the following month the company adopted Vertiglas as a trademark for the thin sheet glass produced by a modification of the Pennvernon process.

Combustion-Engineering Inc. (C-E) was the first company in the sheet glass industry to be granted a float glass license in 1970. C-E had entered the sheet glass and safety glass businesses by acquiring Mississippi Glass Co. in 1968 (Uusitalo 1995b:99) and Hordis in 1969 (Uusitalo 1995b:99), respectively. To replace sheet glass, a full range of commercial thicknesses should be available. In 1970 Pilkington could produce float glass 2 mm thick. In the same year float glass 4 mm thick started to compete with sheet glass in the Canadian market. In February 1971 TGI wrote "keep your eyes on these float glass developments - both in technology and marketing. There may be some surprises ahead." Child (1971b) referred to several industry experts when discussing the matter: Will float replace sheet? One of the experts, a representative of a company which produced both float glass and sheet glass, said: "The substitution of float for plate eliminates the need for grinding and polishing, as well as the people to operate these processes. Window glass, like float, is made as a finished product, and it comes out as a continuous ribbon. However, if window glass manufacturers convert to float, the company is, in effect, adding technology and cost to the window glass process. But, of course, if the superior characteristics of float are desired, this conversion may be justified."

Another industry expert continued that one did not have to do anything to float glass to get the 1/4″ (6.5 mm) thickness since it represented the nominal thickness of float glass. If one was going to make float glass as thin as double-strength (0.18″ or 4.6 mm) or single-strength (0.090″ or 2.3 mm) one had to pull it and add more controls to the process. This was more costly. With the sheet glass process, you could achieve those thicknesses at no extra cost over the sheet glass operation.

Moreover, the position of an independent sheet glass producer would be enhanced if the present day float glass manufacturers (i.e. PPG and LOF) phased out their sheet glass plants. According to this article, Guardian and Ford were marketing float glass at sheet glass prices. The rumors were denied by the companies. In fact, Ford, a third company with both float glass and sheet glass operations in the 1960s, had shut down its sheet glass operations in May 1970. However, at the end of the article Mr. Wardrop, general manager of Ford's glass division noted that anyone who thinks float glass will not compete directly with sheet glass in the future is "quite mistaken." As a conclusion to this article one can say that the replacement of sheet glass with float glass was still controversial in May 1971.

In 1971 Asahi Glass Co. (Asahi), a Japanese company and one of the largest sheet glass manufacturers in the world, announced that it had developed a new vertical drawing process that retains the advantages of Fourcault, while realizing the advantages of the Pittsburgh process (Uusitalo 1995b:101). In October 1971, LOF closed its Shreveport, Louisiana sheet glass plant due to financial losses. The company cited two causes: importation of sheet glass, forcing U.S. plants to close or to operate well below capacity and "unrealistic union wage and benefit demands made on the unprofitable window glass operations" (Uusitalo 1995b:101). Next year a politician from Pennsylvania said that the President Nixon "has driven another nail in the coffin of the domestic glass industry" by his approval of a three-step reduction in import duties for window glass (Uusitalo 1995b:101).

In December 1971, PPG increased the sheet production capacity of its plant in Fresno, California (Uusitalo 1995b:101). At the end of 1972, PPG was still advocating its Pennvernon method. "The basic flat glass product line will consist of float glass, the new standard quality, and sheet glass," Mr. Hainsfurther, vice president of PPG, said in May 1972. He emphasized that sheet glass would continue to be a vital part of the firm's product mix in response to continued strong demand in residential construction and other markets (Anonym 1972:9). The same information was given in the 1972 Annual Report dated February 5, 1973 (PPG 1972:5). By 1972 PPG was the largest sheet glass producer in the United States (AGR, November 1973:9).

In 1974 C-E shut down the Florette sheet glass plant while ASG Ind. closed its Okmulgee, Oklahoma plant. In 1976, PPG closed its Mount Vernon sheet glass plant because the company did not foresee any significant future growth in sheet glass demand due to continuing imports of this glass in substantial quantities and at very low prices, and a recent trend on the part of domestic producers to offer float glass in traditional sheet glass thicknesses (Uusitalo 1995b:101). According to PPG's 1976 Annual Report, the company after having developed its own floatPGG glass, started quickly phase out its existing sheet glass facilities and replacing them with the floatPPG glass lines. By 1980 PPG had converted all its sheet glass production in the United States into floatPPG operations. LOF also closed its last sheet glass plant in Charleston, West Virginia in 1980. In 1977, ASG Ind. shut down its sheet glass plant in Jeannette, Pennsylvania. Next year the company was acquired by Fourco Glass Co. The company thus formed adopted the name AFG Industries (AFG). In 1979 Asahi entered the United States sheet glass business by acquiring the Clarksburg, West Virginia sheet glass plant from Hordis Glass Co. The acquired company took West Virginia Flat Glass as its name and it utilized the Asahi vertical drawing method developed by the parent company. In 1983 Asahi shut down the plant (Mushakoji 1986).

4.4.1.3 Summary of the Events in the United States After the Introduction of Float Glass

The Plate Glass Industry (SIC 32112). Float glass was quickly accepted as a raw material for safety glass. The main customer, the auto industry, was growing rapidly in the 1960s. The increased demand for tempered safety glass (1/4″ or 6.5 mm thickness used in the side and back windows of a car) accelerated the diffusion of Pilkington float glass, the nominal thickness of which just happened to be 1/4″. However, float glass was licensed only to the companies which had been in the plate glass industry before the introduction of float glass. Many 'outsiders' tried to gain access to the capital-intensive plate glass industry via a Pilkington license. St. Gobain's entry into the U.S plate glass industry might also have had an effect on the speed of the diffusion of float glass among the three U.S. manufacturers.

Float glass accelerated globalization of the industry. In the late 1960s, Pilkington built two float glass lines in Canada to produce float glass for the safety glass manufacturers in Canada. Canada preferred local safety glass supplies for its growing auto industry. In 1968, tinting float glass was also accepted quickly in the auto industry. In 1970 St. Gobain, to which Pilkington did not grant a license for the U.S. market, had to withdraw from the U.S. flat glass market. Around 1970 the capacities of float glass lines increased considerably. The thinner thickness became available. Pilkington had to change its licensing policy since pressure from the companies outside the plate glass industry had increased. The first 'outsider' was Guardian Industries, a U.S. based safety glass manufacturer. In the mid-1970s, PPG developed its own float glass process. At the same time the rest of the then operating plate glass lines were shut down.

The Sheet Glass Industry (32111). In the sheet glass industry, imports to the United States had increased since the mid 1950s, accounting for more than 25 % of the domestic demand throughout the 1960s. Increased imports raised tariff controversies, which existed throughout the 1960s and the early 1970s. Three U.S. Presidents, Kennedy, Johnson and Nixon, were involved in the tariff discussions. In the 1960s sheet glass producers still invested in new drawing machines and in new plants. PPG, for instance, built new plants in Fresno, California and in Owens Sound, Ontario, Canada. The U.S. manufacturers PPG and LOF even invested in a sheet glass plant in Europe.

R&D continued active throughout the 1960s. In 1969, PPG introduced Verti-draw sheet glass, the quality of which was said to be equal to that of float glass. In 1970 float glass was introduced in this branch. Although float glass was invading the sheet glass industry, in February 1973 PPG still said that sheet glass was one of their base product lines in the flat glass industry (PPG 1972). However, in the mid-1970s, after having invented its own floatPPG glass, PPG quickly shifted its flat glass production capacity to the floatPPG glass process. At the end of the 1970s sheet glass producers had disappeared with few exceptions. One of them was Japanese Asahi Glass Co., which acquired a local sheet glass plant and modified the old Fourcault machines to their own Asahi method.

4.4.2 Europe

Two large companies, Pilkington from the UK and St. Gobain from France, have had a dominant position in the European flat glass industry (both in sheet and plate glass) throughout the twentieth century. This is why the short histories (ending around 1960) of these two companies are presented. However, an overall view of the flat glass industry in Europe at the beginning of 1960s is given in Table 4.3.

Pilkington Brothers Ltd., the United Kingdom. In 1826 William Pilkington joined with two well-known glassmakers to form the St. Helen's Crown Glass Company and later Pilkington Brothers, Ltd. (1894). Since the mid-1930s, Pilkington has been the only British producer of both plate and sheet glass. Because the plate glass process was so capital-intensive, manufacturing was centralized at Pilkington's original St. Helens location, where coal, limestone, dolomite, alkali, and iron-free sand were also abundant within reasonable distance (Quinn 1977).

In the early 1900s, Pilkington concentrated on plate glass production. In 1923 together with Ford, Pilkington installed the industry's first continuous plate manufacturing process at St. Helens. 12 years later Pilkington introduced a 'twin' machine to grind both sides of a plate glass ribbon simultaneously. In 1929 Pilkington entered the safety glass market by forming with Triplex, the largest British producer, a joint company, Triplex Northern to produce laminated glass. In 1955 Pilkington acquired a 28 % stake in Triplex (Barker 1977a; Quinn 1977).

In 1946 Pilkington had no glass production facilities outside the UK except a partly owned operation in Argentina. In 1951 the company started sheet glass production in Canada and South Africa. In the early 1950s, Alastair Pilkington began to look for easier ways to produce plate glass. After 7 years of secret development, float glass was introduced in January 1959. This invention gave Pilkington a competitive edge in the plate glass industry. Eventually the company decided to license the technology (Barker 1977a, 1994; Quinn 1977:13). Lord Pilkington, the chairman of Pilkington in 1948–1973, described certain key aspects of the resulting strategy as follows (in Quinn 1977:13):

> We had the great benefit of time to decide upon the strategy (the development of float glass took seven years). A great deal was said about ethics: that it was not our job to deliberately deny any existing glass competitor the opportunity of living in competition with us. I don't think we were shortsighted or rapacious… There was a great deal of investment worldwide in plate (glass), and people needed to have time to write off this plant or convert over. The alternative was chaotic disruption of a great industry.

Table 4.3 Major flat glass and safety glass producers in Europe at the beginning of the 1960s

Company	Origin	Co-operation or ownership	Plate	Sheet	Safety
Pilkington	UK		×	×	×
St. Gobain	France		×	×	×
Boussois	France		×	×	×
Glaverbel	Belgium		×	×	×
St. Roch	Belgium	St. Gobain	×	×	×
Delog	W-G			×	×
Detag	W-G			×	×
Vegla	W-G	St. Gobain	×	×	×
S.I.V.	Italy	LOF/US	×	×	×
Pennitalia	Italy	PPG/US			×

St. Gobain, France. The Royal Manufactory of Mirrors, a direct ancestor of the St. Gobain, was created in 1666. The Royal Manufactory subsequently underwent many structural modifications before becoming St. Gobain in 1959. St. Gobain entered the window glass business in 1914 with a license from American Window (a machine blowing system) and later on in the 1920s and the 1930s it started to use both the Fourcault and the Pittsburgh drawing methods (Daviet 1989:108–111).

In 1926 the company began to produce plate glass with a new method, continuous casting, which also included the grinding and polishing units. Around 1927 St. Gobain launched its Sécurit tempered safety glass. In 1959 the company introduced a twin polishing unit, Jusant, in the plate glass industry (Daviet 1989; Hamon 1988). In the early 1960s St. Gobain was a conglomerate company whose main activities were in four industries. Glass was the largest one, accounting for a 59 % share of total turnover. The other industries were chemicals (32 %), petrol (11 %), pulp (5 %), and miscellaneous (2 %) (Saint Gobain 1965).

4.4.2.1 The Plate Glass Industry

In 1962 and 1963, Pilkington started its second and third float lines in Cowley Hill (Grundy 1990:139). The third line, CH4, had produced plate glass before conversion to float glass. The conversion was done by keeping the glass furnace, removing the first 40 m of the lehr, putting a float bath in between the tank and lehr, and then extending and redesigning the lehr (Grundy 1990:45). In Summer 1963 Pilkington's all three of float glass lines had problems. In the first one, massive R&D work was done in order to produce thinner float glass. The other two had quality problems (Grundy 1990:53). 4 years later the company terminated its plate glass production. In the early 1970s, it had four float glass lines in production in Europe. All of them were situated in the UK. In summer 1968 Pilkington started flat glass containerization. The company could export 360,000 tons of glass per week in 20 containers. This special unit had been designed to carry glass to

Holland without the use of pallets or packing cases. Similar projects were being considered for other markets (Uusitalo 1995b:106).

St. Gobain, one of the largest flat glass manufacturers in the world, was at first reluctant to acquire a float glass license. This was due to both the huge investment (1959–1962) in a plate glass plant in the United States and their new twin polishing unit, Jusant. It was very difficult for the company to withdraw from the U.S. plate glass business just after the inauguration. The situation was the same with the new twin polishing units in their Chantereine, Pisa and U.S. plants. The company could not abandon its recently invented machinery (Daviet 1989). In 1967 the company's capacity still consisted of 1,500 tons/day for plate compared with 700 tons/day for float.

St. Gobain's local French competitor, Boussois, a manufacturer of plate, sheet, special glass and glass fiber, received the float glass license for French territory in 1962. Float glass production started in Boussois, France in February 1966. In 1966 Boussois–Souchon–Neuvesel (BSN) was formed in a merger of Boussois Glace and Souchon–Neuvesel, a French glass container manufacturer. Behind this merger was the dynamic managing director of Souchon–Neuvesel, Antoine Riboud. In 1968 Antoine Riboud tried to merge BSN and its larger competitor, St. Gobain, which was four times larger, but he failed. 4 years later, in 1972, BSN acquired Glaverbel. Glaverbel had received the float glass license in 1962 and its float production had started in March 1965 (Hamon 1988; Daviet 1989).

Both French producers had operations in West Germany. St. Gobain had again been involved with its German subsidiary, Vegla, after the end of the Second World War. As mentioned above, St. Gobain had entered the German market via a foreign direct investment (FDI) in 1855. Boussois had acquired the majority share of Delog in 1960. Delog had co-operated for a long time with Detag, a third German flat glass manufacturer. They had cross ownership and a jointly owned subsidiary, Eomag, in Austria. Both of the companies had also produced 'plate glass,' although they used thick sheet glass as their raw material. The companies were not 'pure' enough in the plate glass industry to be accepted as a licensee for Pilkington's float glass. The situation was difficult. Detag had applied in the early 1960s for a license from Pilkington, but without success. Delog could have sublicensed the float process via its French parent company, but then Detag would have been in a worse position. The companies continued co-operation throughout the decade. In the late 1960s, BSN delivered float glass from its French plant to both Delog and Detag.

In February 1970, three companies, Delog, Detag and BSN established a jointly owned company Floatglas GmbH to produce float glass in West Germany. 6 months later, to achieve more synergy, Delog and Detag were merged to form Flachglas AG. In March 1974 Flachglas started its first float glass line. Vegla (St. Gobain) had started float glass production as early as 1966 and in 1974 it had three lines in operation in West Germany (Spoerer 1987). In 1973 Vegla ceased its plate glass production (Friedrich 1974:47).

At the beginning of the 1970s, St. Gobain had become the major holder of Pilkington's float glass license in Europe. The company invested in float glass lines in France, Belgium (via its subsidiary St. Roch), Italy, Spain and West Germany (via

Vegla). In 1974 St. Gobain had nine float glass production units in Europe and its production accounted for one-third of the float glass produced in Europe. The large expansion was economically possible due to the merger of St. Gobain and the financially strong Pont-a-Mousson. This merger took place in 1970.

In 1962, the U.S. based LOF was a minority owner (with Italian investors) in a project to establish Societa Italiana Vetro (S.I.V.), a plate, window and safety glass plant with 2,000 workers in Italy. S.I.V started production in January 1967. In 1974 S.I.V. began float glass production in its facilities. In 1979 Guardian together with a Liechtenstein company, United Export Corp., founded a float glass manufacturing company, Luxguard, in Luxembourg. Guardian had 60 % of the shares and United Export Corp. had the remainder. United Export Corp. was a fabricator and distributor of glass products throughout Europe, the Middle East and Africa.

4.4.2.2 The Sheet Glass Industry

The sheet glass industry was highly fragmented in the early 1960s. There were 35 glass producers in Europe. However, among these 35 firms were four companies St. Gobain, Glaverbel, Boussois and Pilkington, which had either a minority or a majority ownership in 25 companies. The remaining six firms were not financially strong enough to acquire a float glass line. (Derclaye 1982:144). In the early 1960s St. Gobain accounted for only around 2 % of the sheet glass produced in the world. A new sheet glass plant with six drawing machines was installed in Aniche, France in 1962. Pennitalia in Italy (60 % owned by PPG from the U.S) was dedicated in May 1964. The plant had six Pennvernon machines. S.I.V. had also invested in a sheet glass plant in the mid-1960s.

As mentioned earlier the two German companies, Detag and Delog did not produce plate glass. Instead, the firms used thick sheet glass as a raw material for polished plate glass (Friedrich 1974:153). Demand in West Germany for ordinary sheet glass for construction was growing in the early 1960s. Delog, Detag and Vegla, which also had plate glass production, had to invest further in sheet glass plants. Detag started a new Fourcault plant in 1963. In 1957 Delog, which had earlier used both Colburn and Fourcault drawing machines, invested in its first Pittsburgh plant. In 1964–1965 Delog built another Pittsburgh plant with a total of eight drawing machines in Wesel. The old Pittsburgh plant (inaugurated in 1957) was then shut down and reopened in 1969. In the late 1970s Flachglas (formed in the merger of Delog and Detag) shut nine of its sheet glass plants, leaving one for special purposes.

To replace sheet glass, a full range of commercial thicknesses of float glass should be available. In 1969 the commercial range of thicknesses of float available was 3–15 mm (Pilkington 1969a:11). Next year Pilkington could produce float glass of 2 mm thickness. Pilkington's fourth float glass line (CH2 started up in September 1972) was designed to produced 4.0 mm float glass particularly for the European sheet glass markets (discussion with Sir Antony Pilkington, 14 July 1995; Glass, March 1973:96).

Glaverbel, once one of the largest sheet glass producers in the world, had serious troubles with its sheet glass operations. In 1974–1975 only three out of its eight sheet glass plants were in operation, and the three remaining ones were running at below 60 % of capacity (Financial Times, 20 November 1975). In 1980 PPG had its last sheet glass production unit in Salerno, Italy. The plant was converted to floatPPG in 1983. Around 1979/1980, BSN withdrew from the flat glass business. In 1979 the German subsidiary of BSN, Flachglas, was acquired by Pilkington. The following year, the company sold the Belgian and Dutch subsidiaries, Glaverbel and de Maas, to the Japanese Asahi and the French operations in Boussois to the U.S.-based PPG. In 1983 Asahi converted the sheet glass operations in Holland to the float glass process.

4.4.2.3 Summary of the Events in Europe

The Plate Glass Industry. In 1962–1963 Pilkington built its second and third float glass lines. In 1967 the company closed its last plate glass line. Other European plate glass companies were at first reluctant to switch to float glass. Glaverbel of Belgium was the first to start the float glass production in March 1965. However, later on, in the late 1960s and in the early 1970s, St. Gobain invested in nine float glass lines. Concentration in the plate glass market took place. A French company, BSN, acquired the Belgian manufacturer Glaverbel in 1972. In 1979 BSN withdrew from the flat glass industry by selling its French operation to the U.S. based PPG, its German operations to Pilkington, and its Belgian operations (Glaverbel) to the Japanese Asahi. At the same time Guardian built a float glass line in Luxembourg.

The Sheet Glass Industry. In the mid-1960s the U.S. producers PPG and LOF invested in the sheet glass industry in Europe. The U.S. market was a good export area for European producers throughout the 1960s. In 1970 Detag, the last independent flat glass producer, was sold to BSN. After the introduction of float glass in the UK in 1959 it took 8 years in France and West-Germany, seven in Italy and 18 in Sweden to introduce it. By 1981 all these countries except Italy used 100 % of float glass. In Italy the percentage was 68 (Ray 1984:81).

References

Allen AC (1967) One of the world's largest glass tanks on stream. Ceramic Industry Magazine, October, pp 50–51

Allen AC (1968) New Canadian plant draws 14 Miles of sheet glass per day. Ceramic Industry Magazine, December, pp 52–54

Anderson P (1988) On the nature of technological progress and industrial dynamics. Unpublished Ph.D. dissertation, Columbia University

Anonym (1964) First U.S. float glass plant now on stream, The Glass Industry, April, pp 176–178

Anonym (1968) Tinting Float Glass within production line, The Glass Industry, January, pp 20–21

Anonym (1972) PPG builds world's largest float plant, The Glass Industry, May, pp 8–10, 13

Arbose J (1986) Why once-dingy Pilkington has now that certain sparkle. Int Manage 41(8):44–45, 48

Barbour E (1971) Pilkington float glass (A), HBS case services. Harvard Business School, 9-672-069 rev. 2/82

Barker RF (1964) Report to the U.S. Tariff Commission. The Glass Industry, August, pp 426–427, 444

Barker TC (1977a) The glassmakers. Pilkington: the rise of an international company 1826–1976. Weidenfeld and Nicholson, London

Barker TC (1977b) Business implications of technical development in the glass industry, 1945–1965: a case-study. In: Supple B (ed) Essay in British business history. Claredon Press, Oxford, pp 187–204

Barker TC (1986) Pilkington, the reluctant multinational. In: Jones G (ed) British multinationals: origins, management and performance. Aldershot, Cambridge, pp 184–201

Barker TC (1994) An age of glass. Pilkington: the illustrated history. Boxtree, London

Caulkin S (1987) Pilkington after BTR. Management Today, June, pp 43–49, 127, 129

Child FS (1971) Turmoil in the flat glass industry. The Glass Industry May, pp 167–169

Cover Story (1962) New contender for the plate glass market. The Glass Industry November, pp 597–601

Daviet J-P (1989) Une multinational a la Française. Histoire de Saint-Gobain 1665–1989. Fayard, Paris

Derclaye M (1982) La diffusion d'une innovation technique, le "float-glass" et ses consequences sur le marche du verre plat. Annales de Sciences Economiques Appliquees 38(1):133–148

Doyle PJ (ed) (1979) Glass making today. Portcullis Press Ltd, Redhill

Editorial (1967) About the Tariff, The Glass Industry, February, p 114

F.G.M.A. Convention Report (1967) Kennedy round results denounced. Glass Digest, September, pp 62–65, 76

Frederiksen PC (1974) Prospects of competition from abroad in major manufacturing oligopolies: case studies of flat glass, primary aluminum, typewriters, and wheel tractors. Unpublished Ph.D. dissertation, Washington State University

Friedrich H (1975) Der technische Fortschritt in der Glaserzeugung. Eine Untersuchung über die Auswirkung des technischen Fortschritt auf den Strukturwandel in der Flachglasindustrie (Bochumer Wirtschaftswissenschaftliche Studien Nr. 7), Bochum

Grundy T (1990) The global miracle of float glass: a tribute to St. Helens and its glass workers. St. Helens, Merseyside

Hamon M (1988) Du Soleil a la Terre, Une Histoire de Saint-Gobain. Jean-Claude Lattès, Paris

Kinkead G (1982) The raging bull of glassmaking. Fortune, 5 April, pp 58–64

McCauley RA (1980) Float glass production: Pilkington vs. PPG, a comparison of the two processes—their similarities and differences. The Glass Industry, April, pp 18–22

Mushakoji K (1986) The process of internationalization at Asahi Glass. Int Manage March:73–74, 79–80

Perry RC (1984) Float glass process: a new method or an extension of previous ones? The Glass Industry, February, pp 17–19, 31

Persson R (1969) Flat glass technology. Butterworths, London

Phoenix Award Winner (1981) Sir Alastair Pilkington. The Glass Infustry, July, pp 15–17

Pilkington A (1963) The development of float glass. The Glass Industry, February, 80–81, 100–102

Pilkington A (1969a) The float glass process. Proc R Soc London, A 314:1–25

PPG (1972) Annual Report

Quinn JB (1977) Pilkington Brothers LTD. The Amos Tuck School of Business Administration, Dartmouth College

Quinn JB (1979) Technological Innovation, entrepreneurship, and strategy. Sloan Manage Rev Spring:19–30

Ray GF (1984) The diffusion of mature technologies. Cambridge University Press, Cambridge

Saint-Gobain (1965) Compagnie de Saint-Gobain, 1665–1965. Livre du tricentenaire, Paris

Salmans S (1980) Pilkington's progressive shift. Management Today, September, pp 63, 66–73

Sheet glass and the tariff (1962) The Glass Industry, pp 240–241, 281

Simpson HE (1961) The glass industry—1960 a review. The Glass Industry, January, pp 67–70

Simpson HE (1963) The glass industry—1962 a review. The Glass Industry, January, pp 71–74

Skeddle RW (1980) Expected and emerging actual results of a major technological innovation—float glass. Omega 8(5):553–567

Spoerer M, Busi A, Krewinkel HW (1987) 500 Jahre Flachglas, 1487–1987 Von der Waldhütte zum Konzern, (in German, 500 Year Old Flachglas, 1487–1987, from a small business to a concern). Karl Hofmann Verlag, Schorndorf

Takahashi S, Ichinose M (1980) New vertical draw process for sheet glass. The Glass Industry, April, pp 24, 29-30, 32

The Tariff Situation (1961) The Glass Industry, 429–432, 469

United States Tariff Commission (1968) sheet glass (blown or drawn flat glass). Report to President (No. TEA-IR-7-68) TC Publication 262, Washington D.C. September 1968

Uusitalo O (1995a) The flat industry—the effects of the float glass on the industry structure. Licentiate thesis B-156, Helsinki School of Economics, Helsinki

Uusitalo O (1995b) A revolutionary dominant design—the float glass innovation in the flat glass industry. Dissertation A-108, Helsinki School of Economics, Helsinki

Uusitalo O, Möller KEK (1997) Macro network dynamics—the evolution of the flat glass industry. In: Gemunden HG, Ritter T, Walter A (eds) Relationships and networks in international markets. Pergamon, New York, pp 411–426

Vincent GL (1960) PPG operates largest, most modern U.S. plate glass plant. Ceramic Industry Magazine, September, pp 100–105, 136

Vincent GL (1962) How Ford produces plate in world's largest integrated glass plant. Ceramic Industry Magazine, November, pp 50–55, 80

Wierzynski GH (1968) The eccentric lords of float. Fortune, July, pp 90–92, 121–124

Chapter 5
Findings in Anderson's (1988) Study

Abstract This chapter has five sections. In the first section, the empirical findings from the U.S. flat glass industry in 1920–1980 of the original study of cyclical model of technological change are revealed and discussed. In the original study used four-digit SIC codes for the definition of the industry. The focus in this research is to find out the impacts of this definition on the depth of the analysis. The second, third and fourth sections redefine the performance parameter, technological discontinuity and dominant design concepts of the cyclical model of technology, respectively. In the fifth section the flat glass industry is tested by means of a modified version of the cyclical model of technological change. For the test the U.S. plate glass and sheet glass industries are defined with more accurate five-digit SIC codes. Contrary to the original study of the cyclical model of technological change float glass emerged as the dominant design in both the plate glass industry and plate glass industry both in the United States and Europe. Finally there are critical comments on the entirely quantitative research method and the use of four digit SIC codes. These misinterpretations of the industry are discussed at the beginning of the chapter.

Keywords Dominant design · Performance parameter · Risk of quantitative method

This chapter has five sections. In Sect. 5.1, I reveal Anderson's (1988) empirical findings from the U.S. flat glass industry in 1920–1980 in his study for creating the cyclical model of technological change. I use only Anderson's (1988) findings since Anderson and Tushman (1990, 1991) are based on Anderson's work. As was mentioned Anderson used four-digit SIC codes in defining the industry. My focus is the find out what are the impacts of that definition on the depth of the analysis. Sections. 5.2, 5.3 and 5.4 redefine the performance parameter, technological discontinuity and dominant design, respectively. In Sect. 5.5 the flat glass industry is tested by means of a modified version of the cyclical model of technological change (Anderson 1988). In the test I define the U.S. plate glass and sheet glass

O. Uusitalo, *Float Glass Innovation in the Flat Glass Industry*,
SpringerBriefs in Applied Sciences and Technology,
DOI: 10.1007/978-3-319-06829-9_5, © The Author(s) 2014

industries with five-digit SIC codes. I also test whether the modification of Anderson's (1988) model and the definition of the industry influence the emergence of float glass emerged as the dominant design in the United States and Europe. This chapter also comments on the entirely quantitative research method.

5.1 Anderson's (1988) View of the U.S. Flat Glass Industry in 1920–1980

This section selects some of Anderson's (1988) interpretations from the events in the flat glass industry in 1920–1980. These quotations which are mainly from "Appendix C: The U.S. Flat Glass Industry" give his view of the U.S. flat glass markets in 1920–1980. Only the first quotation is from Chapter 5, Empirical Results—Technological Change (Anderson 1988:133). All the headings are from Anderson's work. The direct quotations are used to preserve Anderson's original intentions. Each quotation is followed by immediate comments from the point of the number (either four or five) of digits in Standard Industry Classification (SIC) code.

CHAPTER 5 EMPIRICAL RESULTS– TECHNOLOGICAL CHANGE
INTRODUCTION
Even if we limit our survey to the twentieth century, we observe some 220 years of glass industry history (collapsing flat glass into one industry after 1960),…. (1988:133).

APPENDIX C: THE U.S. FLAT GLASS INDUSTRY
THE TECHNOLOGY OF FLAT GLASS MANUFACTURE

Figure C-1 (Anderson 1988:374) STATE OF THE ART SQ. FT. PER HOUR: US WINDOW GLASS MACHINES ILLUSTRATES four technological discontinuities: Lubbers machine cylinder (in 1903), Improved Lubbers (1907), Colburn machine glass drawing (1917) and float glass (1963) in the U.S. window glass industry.

For most of its history, plate glass was ground and polished to make both sides perfectly parallel and free from optical distortion. The grinding process was quite labor- and equipment-intensive, and differentiated plate glass from ordinary window glass. Inexpensive, lower quality window glass suited the residential housing market, while higher quality, expensive plate glass was used in commercial buildings and vehicles. As will be discussed, the float glass has all but eliminated grinding and polishing, and there is now no technical distinction between plate and window glass (1988:335).
For most of the industry's history, plate glass and window glass firms utilized different technologies, and did not compete directly with one another. Accordingly, the discussion in this chapter will address the evolution of window glass manufacture first, then the evolution of plate glass manufacture, until the two stories are united by the float glass revolution (1988:336).

According to the first quotation and Figure C-1 (in Anderson 1988, Appendix C, which shows that float glass was introduced in the window glass industry in 1963) float glass immediately converged the plate glass and the sheet glass industries into a single industry called the flat glass industry. This indicates that the

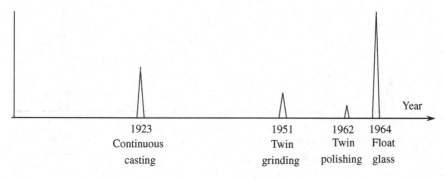

Fig. 5.1 Technical discontinuities in the U.S. plate glass processes

use of four-digit SIC codes were accurate enough. However, the last two quotations from Anderson's (1988) study tell that the convergence was gradual and, thus ask for separate study for both window glass and plate glass industries. However, this would have required the use of five-digit SIC codes in the analysis.

THE EVOLUTION OF FLAT GLASS TECHNOLOGY

Plate glass

The progress of plate glass machinery is not as well document as that of window-glass machinery. Since no data exists allowing us to assess quantitatively the impact of the improvements noted above, we cannot construct a diagram such as figure C-1 to illustrate the technological progress of plate glass manufacture.

If one uses five digits SIC codes the technological progress in the U.S. plate glass industry could be construct. There were continuous casting in 1923 and float glass as will be shown in Fig. 5.1.

The Float Glass Era

The effect of float glass was twofold. First, it greatly reduced costs by eliminating the time-consuming and expensive process of polishing plate glass, and speeding up the rate of production dramatically. Second, over time it eliminated the historic division of the flat glass into window and plate branches. This division into non-competing sectors occurred not only because the two products addressed different markets, but because their manufacture employed different production methods polishing plate glass demanded expertise and equipment not possessed by the window manufacturers.

The float glass process was immediately adopted by the plate glass producers, licensed by Pilkington, and for a time was used only for plate glass because the float process was more expensive than simple drawing. Over time, however, the large economies of scale achieved by float glass producers permitted them to lower costs near that of window glass producers, and since drawn glass was inferior quality, window glass manufacturers were forced to adopt the float process. Ultimately, the distinction between window and plate glass producers disappeared, and both periodicals and directories began referring to the surviving producers as 'flat glass industry (Anderson 1988:350).

According to the above excerpts, Anderson seemed to have understood the nature of the gradual, over time, convergence of the two subindustries and the rapid acceptance (without any licensing problems) by the plate glass industry. The excerpts

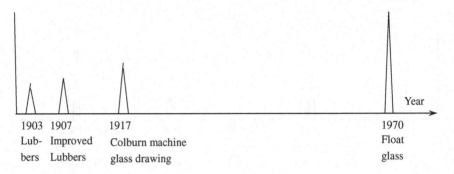

Fig. 5.2 Technological discontinuities in the U.S. sheet glass processes

also report about the quality and price differences of the two products. However, it does not report anything of the different thicknesses of sheet glass and plate glass.

> The Pilkington process diffused relatively slowly, due to the long time required to build a float glass factory (up to four years)..... Between 1964 and 1972, 12 float glass plants were constructed, while ten plate and two sheet glass facilities employing drawing technology were dismantled. By the late 1960s, all new flat glass plants used the float process, and by 1980 practically all U.S. production employed this method (1988:350–351).

Earlier Anderson said that the float glass process was immediately adopted by the plate glass industry instead of diffusing slowly. It took quite a long time, three and a half years, before the first license was granted to a U.S.-based plate glass producer (Barker 1994:81), but once the license was granted, the U.S. plate glass producers set records (as mentioned in the U.S. case) in converting their existing plate glass lines to float glass or in erecting totally new lines (Derclaye 1982:146).

> *Dominant design*
> Again, patent considerations precluded the emergence of a dominant design. The Pilkington process has been subjected to a number of small improvements, whose effect is noted in figure C-1. However, no 'improved version' of the design has emerged as a standard expression of the process. Any firm which wants to build a float glass factory must apply for a Pilkington license (except for PPG, with its own patented process). The licensee installs the standard Pilkington line, incorporating the latest incremental improvements which have accrued over the years (Anderson 1988:351).

The capacities of the first float glass lines installed in the U.S were identical. In Figure C-1 Anderson says that float glass was introduced in the window glass industry in 1963 (compared with 1970, as will be shown in Fig. 5.2), which means that the convergence of the subindustries was immediate. At the end of the above quotation, Anderson talks of a standard Pilkington line, which, however, did not emerged as the dominant design. PPG in fact introduced its own float glass process in 1975, 11 years later than the inauguration of its first float glass line in Cumberland, Maryland. The first lines worked under Pilkington's license. Anderson also admits this elsewhere (1988:354–355): "Pittsburg Plate, Libbey-Owens-Ford (the Ford is unrelated to the automobile firm) and the Ford Motor Company were the first three adopters of the new technology [float glass]".

Effect on competence

Float glass rendered obsolescent the large investment that firms had in drawing and polishing machinery. To remain competitive, producers eventually had to license the Pilkington technology, and scrap their old forming devices. The competence that plate glass producers had acquired in grinding, which had long constituted a barrier to entry for both producers and crossovers from window glass manufacture, was made irrelevant. Window glass makers found their markets invaded by cheaper glass of quality that could not be matched by drawing machines (1988:351–352).

Anderson did not recognize that Pilkington's patent policy was, in fact, acting as a new entry barrier (instead of the 'grinding and polishing barrier') to the plate glass industry. In 1970, 6 years after the start-up of the first float line in the United States, Pilkington had to change its patent policy since Guardian, a U.S based safety glass company, erected its own float line without a license from Pilkington (Uusitalo 1995b:127). Here again a solid and separate analysis of both the plate glass and sheet glass industries with five-digit SIC codes would have helped in the analysis. First, in the plate glass industry the real entry barrier, Pilkington's licensing policy, would have been seen, and second the international context, huge imports of sheet glass, would have been seen in the sheet glass industry.

PIONEERS OF DISCONTINUITIES AND STANDARDS
Pioneers of discontinuities

Four of the first five firms licensed by Pilkington were dominant figures in the plate glass or window glass industry. Pittsburg Plate, Libbey-Owens-Ford (the Ford is unrelated to the automobile firm) and the Ford Motor Company were the first three adopters of the new technology, though they had to abandon millions invested in continuous-formation plate glass equipment. The next licensee was a maverick; Guardian Industries used the float glass process to diversify into flat glass manufacture. The fifth licensee was ASG Industries, the descendant of American Window Glass, and one of the largest window glass manufacturers. ASG, however, was nearly a decade late in adopting float glass technology, and it did so only after the American firm was acquired by St. Gobain of France, a Pilkington licensee in Europe (1988:354–355).

Guardian Industries was a safety glass manufacturer, which due to high price of float glass made a strategic decision to integrate backwards to the flat glass manufacture. In fact, Guardian Industries started float glass manufacturing without any license from Pilkington. The company hired skilful personnel from the Ford Motor Company. (Kinkead 1983).

St. Gobain entered the U.S. flat glass manufacturing business in the 1920s. In 1958 it merged two firms, Blue Ridge Glass and American Window Glass, to form American St. Gobain (ASG). In 1962 ASG started plate glass production as was mentioned earlier. The U.S. plate glass line was a burden for the parent company (St. Gobain) throughout the 1960s. In 1970 St. Gobain withdrew from the U.S. flat glass market. A group of American investors acquired American St. Gobain and the new firm adopted the initials as its name, ASG Industries (ASG Ind.). ASG Ind. started float glass production in 1973 after St. Gobain had left the company (not after St. Gobain had acquired ASG Ind.). Again a very careful analysis of the plate glass industry with five-digit SIC code would have revealed the strategic move of the French parent company into the U.S. plate glass industry. Pilkington due to its strict licensing policy did not grant a license to St. Gobain in the United States. (Daviet 1989 and Saint-Gobain 1965)

ECONOMIC HISTORY OF THE INDUSTRY
  Window glass
  The float glass era
    The impact of float glass on the window glass manufacturers was one of boom and then bust......Sales in real dollars soared briefly in the late 1950s, dropped for two years, then rose again until the mid-1960s. The increase from 1963-1965 is largely attributable to new production from LOF and PPG float plants, whose output consisted of both 'plate' and 'window' glass under the census definitions". Window glass sales peaked in 1965, and began a long decline that would never be arrested......Much of the reduction can be attributed to the invasion of traditional window glass markets by float glass producers (1988:361).

In the above quotation, Anderson says that the plate glass and window glass subindustries converged when float was introduced in the U.S. plate glass industry. Moreover, Anderson implies that the U.S. manufacturers, LOF and PPG, were, in fact, manipulating the Census Bureau (i.e. writing float glass both under the plate glass and window glass digits). The Census Bureau had realized that float glass had been invented in the plate glass industry; thus the bureau wrote float glass under the plate glass digit. The Census Bureau stopped distinguishing the two subindustries in 1977 since all float glass (both by the Pilkington and PPG processes) produced had accounted for more than 85 % of the total flat glass production. Subsequently, all flat glass was written under the flat glass digit (Uusitalo 1995b:129). Again a separate analysis with five digits SIC codes would have been helpful.

As can be seen from the review of the evolution of the U.S. flat glass market, both LOF and PPG continued their sheet glass production after the introduction of float glass in the plate glass industry. Float glass was not introduced in the window glass industry in the United States in 1963 as Anderson (1988:374, Figure C-1) tells us. The increase in window glass production in 1963–1965 was due to the tariff increase on imported window glass. Window glass imports rose in the latter half of the 1960s, thus reducing domestic production. Float glass started its invasion of the window glass industry in the early 1970s. A solid and separate analysis with five-digit SIC codes would have revealed the industry specific events.

  Plate glass
    PPG and LOF both licensed the Pilkington process in 1959. In the short run, US plate glass sales dropped, partially due to competition from European plate glass manufacturers who had adopted the new technology before the Americans (1988:364).

PPG made the first U.S. licensing agreement in June 1962 (Barker 1994:81) and in general the U.S. companies adopted float glass faster than their European competitors (Derclaye 1982:145–6).

ENTRIES AND EXITS
Window glass
The float glass revolution had little effect because the three dominant window glass producers adopted the new technology, while the remaining smaller producers appeared able to defend their small, niche markets as they had for decades against their larger rivals. (1988:367).

The use if four-digit SIC code, that is concentrating on the flat glass, made Anderson think that the U.S. companies, such as PPG and LOF, having both plate glass and sheet glass manufacturing immediately converged all their production to float glass. This was not the case. As we saw PPG for instance installed sheet glass plants in the late 1960s. In 1972 the company told that sheet glass will be kept as one of their main products.

Anderson (1988) defined the industry demand both in the plate glass and the sheet glass subindustries from the digits of the U.S. Census of Manufacturers (Anderson 1988:122), thus neglecting the imports. He counts the number of entries and exits in his hypotheses. Imports of window glass to the United States met about one-fourth of the demand and equaled one-third of domestic production in the 1960s. Float glass is a technological change in the flat glass industry, which is included in Anderson's work. One cannot say how many of the exits from the window glass industry took place because of the new technology or because of the price competition with imports.

Plate glass
Two companies – Guardian Industries and AFG – did license the Pilkington process in the 1970s as a means of diversifying into flat glass production Anderson (1988:370).

Anderson does not mention the entry of a French manufacturer St. Gobain via its U.S. subsidiary, American St. Gobain (ASG), to the U.S. plate glass industry in 1962. As was mentioned Pilkington did not license its French competitor float glass in the U.S. territory. Thus ASG after having been sold to local investors was able to license float glass. After the sale the company took the name ASG Industries. Later on Fourco Glass Co acquired ASG Ind. in the mid 1978 and picked the name AFG Industries (AFG). Thus, neither ASG nor Fourco Glass Co was diversifying into flat glass production. Again a solid and separate analysis under five-digit SICs codes would have helped.

At least two trade journals were consulted for each industry. Every issue of each journal was scanned. Articles and notices that documented technical changes, entries, exists, mergers, and other relevant information were photocopied, and the information was later transferred to a computerized database (Anderson 1988:117).

The fourth section of each appendix then relates a brief economic history of the industry highlighting surges and drops in demand, major market changes, key legal regulatory events... (1988:219).

According to the previous two quotations, Anderson said that he scanned every issue of each trade journal and that the economic histories point out major events in the industries. However, Anderson did not mention imports, tariffs or dumping investigations in his economic history of the window glass industry in the float glass era (Anderson 1988:361), although dozens of pages were devoted to these topics in TGI and even three U.S. Presidents (Kennedy, Johnson and Nixon) were involved. The Glass Industry trade journal is included in Anderson's references (Table 4-1, p 131). The four-digit (to the level of flat glass) definition of the industry disintegrated him from the sheet glass industry and its international context.

On the one hand, Anderson's (1988) analysis of the U.S. flat glass industry, states that the convergence of the plate glass and the sheet glass subindustries took place over time (1988:350) while, on the other hand, he says that they (the two subindustries) collapsed into one industry after 1960 (1988:133). In Anderson's analysis the convergence of the two subindustries seems to be problematic. The reason may be the use of four-digit SIC codes in defining the industry.

Safety glass industry

Since Anderson defined the industry by four-digit SIC code so pertinent industry, the safety glass industry (SIC 32113), for the plate glass manufacturers was disintegrated from his analysis. The following quotation confirms this:

> The discussion in this chapter will be limited to firms which produce flat glass, not those which produce products based on flat glass Anderson (1988:335).

Thus, he does not illustrate anything about the links from the plate glass industry to the growing car industry via safety glass industry while float glass was introduced to the U.S. plate glass industry.

5.1.1 The Definition of the Industry in Anderson Study

There are some traps, even dangerous ones, if a researcher defines too narrowly the industry. A researcher can easily get on the wrong track when interpreting the empirical material. Once on the wrong track the remainder of the empirical material is viewed from the same vantage point. Rosenberg points out that, in the study of technology, one should avoid too narrow a focus. He argues that:

> ...it is absolutely essential not to develop too narrow a focus in the study of technology, because a narrow focus severs the links between a given technology and many of the factors that will, inevitable, determine its effectiveness and significance (Rosenberg 1982:56).

The research design with the definition of the industry with four-digit SIC code in Anderson's study seems to give a relatively limited view of the flat glass industry. For instance, missing the nature of the gradual convergence of the plate glass and the sheet glass industries may have been one consequence.

Census figures (manufacturing) were used intensively in the analysis of the flat glass industry. The convergence of the plate glass and the sheet glass industries was probably noticed when the census figures were combined in 1977. This convergence was, on the one hand, said to have occurred just after the introduction of float, but on the other hand, it was said to have happened over time. Also the flat (both plate and sheet) glass sales, i.e. demand figures, were compiled from the U.S. Census of Manufactures although a vast amounts of sheet glass was imported. Tariff controversies were ignored in the economic history of the window glass industry. This can be understood since census figures do not include any qualitative data. The quantitative research design based on manufacturing census figures also leaves out strategic foreign direct investments. In 1962 a newcomer into the plate glass industry was noticed in Figure C-11 (Anderson 1988:384) from the census figures, but it was regarded as a domestic entrance. In fact, the plate glass plant was St. Gobain's, a European glass giant's strategic move under its subsidiary, American St. Gobain, to the U.S. plate glass market.

Anderson's research design also seems to be history-driven. He starts his study (see Introduction, p 1) with a history of an entrepreneur from the nineteenth century. In his attempt to obtain help from industry experts, Anderson started interviewing about events from the nineteenth century (rotary kiln from 1895, Anderson 1988:110). It would have certainly been much better to start interviewing industry experts about the most recent events in each industry. In the analysis of the flat glass industry the industry histories were old (flat glass, Anderson 1988:120). He also included the story of crown glass in the analysis of the flat glass industry (Anderson 1988:337). This invention of crown glass later confused the readers, since it is included in the Anderson and Tushman (1991) article, but excluded from the Anderson and Tushman (1990) article. An observant reader will notice that the number of technological discontinuities is 17 in the first article and 16 in the second article.

The quantitative study with a small sample size (16 in Anderson's case) demands a thorough understanding of the events in the research phenomenon. This task is much more risky if neither data nor information on the business or business strategy is collected. In Anderson's case this seems to be so, since there are no references to any business magazine (flat glass) and there were no interviews. The use of four digit SIC (Standard Industry Codes) codes (i.e. safety glass was left out) does not help in this kind of situation, because the strong link between flat and safety glass is neglected.

This is in accord with Miller and Friesen (1982) who say that one of the dangers of longitudinal, narrowly focused quantitative studies of multiple organization (Anderson's study can be regarded to this type) is the omitting of other factors which drive or intervene between dependent and independent variables (1982:1018). They further argue that the researcher in this type of study is usually unable to get close enough to organizations to discover why things happened and just what their consequences were (1982:1019).

5.2 Redefinition of the Performance Parameter

This section discusses and redefines the performance parameter of the dominant design with more specific subindustry related data. The definition of the subindustries with the five-digit SIC codes provide more accurate data. Performance parameter is discussed before technological discontinuity, since the conclusions made in this section will be used in the analysis of technological discontinuity. Anderson used production volume (square feet of flat glass produced in the plate glass and sheet glass industries) as the performance parameter or dimension of merit. In his analyses of process industries, Anderson argues for his choice of performance parameter as follows:

> In the first four industries (the cement, container glass, plate glass and sheet glass manufacture) the output is essentially a commodity, and since scale economies are important, the key determinant of production cost is the output per unit time of the most productive process equipment in existence (Anderson 1988:105).

Tushman and Romanelli (1985) discuss the effectiveness and the efficiency of an organization, and they say by quoting Katz and Kahn (1966):

> Organizations must be effective in producing a product or service that is desired by an external economy such that a flow of resources is maintained, and efficient with respect to internal resource utilization (Tushman and Romanelli 1985:174 and Table 1 in p 175).

In other words, effectiveness is external to the organization, and efficiency is internal to the organization. It seems that Anderson's definition presumes that efficiency is equal to effectiveness, since his definition says that each type of production technology in given industries has equal product attributes (for instance in the flat glass industry: thickness and quality) and equal and constant cost structures. The equal cost structure means that a new process technology cannot have lower variable and/or fixed costs.

Anderson used square-feet-per-hour capacity as a performance parameter or a 'dimension of merit' for a flat glass forming machine or process of plate glass and window glass. Anderson (1988) said nothing about the thickness of the flat glass, which has an effect on the capacity of a forming machine or process. People in the U.S. sheet glass branch talk about single strength (0.090" or 2.3 mm) and the equivalent. The thickness (Perry 1984) and the price of float glass were two major variables affecting the entrance of float glass into the sheet glass industry. It would have been better for Anderson to compare the costs (the sum of variable and fixed costs) instead of the capacity of a glass forming machine or process. In fact, production costs were affected by all the discontinuities in the sheet glass industry studied by Anderson (1988). The Lubbers machine (i.e., mechanized blowing) eliminated the job of many hand-blowers because of lower costs. The situation was also similar in drawing machines, which swept away the rest of the hand-blowers and the machine-blowing process.

In analyzing the diffusion of float glass, Derclaye (1982) used the variable and fixed costs per ton of flat glass produced as a performance measure. According to him, the variable and fixed costs of float glass were respectively 15–30 and 25–50 % less than those of plate glass when float glass was introduced. Capacities of the production lines are expressed in the plate glass industry either in million square feet of 1/4" thickness glass or tons per day of melted glass. The capacity of the last plate glass line ever built (ASG's line, Greenland, Tennessee, started in

1962 with investment costs of $50 million) was 40 million square feet per year. PPG finished the installation of its first float glass line in the early 1964 alongside one of its plate glass lines. Both of the lines (plate glass and float glass) had a capacity of 50 million square feet per year or 300 tons per day. The investment costs of a float glass line were reported to be $18–25 million in the mid 1960s.

The case description above shows that the main performance parameter is the cost of a square foot or square meter of the flat glass produced instead of the capacity of the process. The annual capacity (nearly 50 million square feet) of the first float glass line (PPG, Cumberland, Maryland) in the United States was the same as that of the plate glass line on the same site. In 1964, Mr. Barker, the vice-president of PPG, stated that the operating costs of the float glass plant had proven to be 25 % less than those of a plate glass plant—largely because float glass did not have to be ground and polished on both sides (Uusitalo 1995b:135). In the present study the redefined performance parameter is the cost of one square foot or square meter of the chosen thickness produced flat glass derived from the costs of one square foot or square meter of the respective flat glass.

5.3 Redefinition of the Technological Discontinuity

This section discusses broadly the definition of technological discontinuity in Anderson's study. According to Anderson, his dominant design model relies heavily on the concept of technological discontinuity (1988:17). The discontinuity between major and minor innovations is common, but the basis of the distinction often varies from study to study. Arrow (1962), for instance, defines a major innovation as one which pulls the prices of a product below its previous perfectly competitive price; a minor innovation is one which does not. According to Anderson, Arrow's definition ignores the performance dimension and does not differentiate between innovations which affect the competitive price a great deal and those which reduce it by a small amount (Anderson 1988:18).

Anderson (1988:19) refers to Tushman and Anderson (1986) when he discusses how a technological discontinuity pushes forward the best cost/performance combination attainable. Tushman and Anderson (1986:439) report their findings in two industries (the minicomputer and the cement), which Anderson (1988) also studied. In Tushman and Anderson (1986) the definition of technological discontinuity is not entirely clear. The process discontinuities are reflected either in process substitution or in process innovations that result in radical improvements in industry-specific 'dimensions of merit' (1986:441). The authors concluded (1986:447) that the technological discontinuities were relatively easy to identify because a few innovations advanced the state of the art so much that they stand out from the less dramatic improvements. However, later on, Anderson (1988:20) admits the difficulties of distinguishing the incremental innovations from the radical ones. Ultimately his focus is on one or two key parameters, 'dimensions of merit,' central to characterizing the cost/performance of an industry's product or process (1988:20). The definitions seem to be the same in both studies (Tushman and Anderson 1986; Anderson 1988).

However, the results of these studies are different. Tushman and Anderson (1986:450) named three major technological discontinuities and Anderson (1988:269) mentioned six. Anderson has added the continuous vertical kiln, the Atlas kiln and the computerized process control kiln. It seems that Tushman and Anderson (1986) regarded the real Atlas kiln as an Edison kiln. Tushman and Anderson (1986) did not recognize the real Edison kiln. Then Anderson split the Dundee kiln into two technological discontinuities: Computer Process Control kilns and a Dundee kiln.

In the cement industry Anderson has a seventh technological discontinuity, suspension preheating (1988:239-243), which was not recognized in Tushman and Anderson (1986). To recognize this technological discontinuity and still prevailing technology (suspension preheating with precalcining) in the cement industry Anderson (1988) had to change the definition of the performance parameter. Anderson called this new dimension of merit a "hidden parameter" (Anderson 1988:106). The dimension of merit, the barrels-per-day measurement of cement production which, in fact, measures the internal efficiency of an organization, was changed to a performance parameter, the energy cost per barrel of cement measuring the external effectiveness. This is, in fact, the problem indicated in the previous section.

Anderson and Tushman later (1990:606) defined technological discontinuity as follows: At rare and irregular intervals in every industry, innovations appear that "command a decisive cost or quality advantage and that strike not at the margins of the profits and the outputs of existing firms, but at their foundations and their very lives" (Schumpeter in Anderson and Tushman 1990). That kind of innovation departs dramatically from the norm of continuous incremental innovation characterizing product classes and may be defined as technological discontinuities (Anderson and Tushman 1990). Process discontinuities are fundamentally different ways of making a product that are reflected in order-of-magnitude improvements in the cost or quality of the product (Anderson and Tushman 1990).

Anderson says that technological discontinuities are identified as tremendous advances in the best performance achievable in the industry at a given time (Anderson 1988:34). Anderson mentioned two cases where his ideas are opposite to those of Foster (1986) (later to be called distinctions A or B). According to Anderson:

A. A technological discontinuity can also take place in the same technological S-curve.
B. A switch from the former S-curve to the later one is regarded as a technological change only if the performance in that given time is increased significantly (Anderson 1988:34).

The first distinction with Foster's (1986) definitions of technological discontinuity seem to accept economies of scale as a technological discontinuity, because it is difficult to imagine any other change in the same technology S-curve than economies of scale. However, later on Anderson (1988:104) says that discontinuous advances which are applications of known technology on an enormous scale would not count as technological discontinuities. These definitions seem to be problematic. This inconsistency can also be seen in the discussion of commercial jets. In Tushman and Anderson (1986:453) Boeing 747 was regarded as a key technological discontinuity, while later on Anderson says: "For instance, wide-bodied jet aircraft effected a quantum improvement over cost/performance capabilities of the first generation jets, though the 747 was a logical extension of its predecessors, and thus perhaps not very novel" (Anderson 1988:19). Anderson, in fact, regarded the 747 only as an 'enormous scale' which would not count as a technological change. Mowery and Rosenberg (1982) also regard the 747 as an 'enormous scale' when they talk about stretching of the fuselage and thus, according to them, modern aircraft are designed so as to develop and exploit technological trajectories (Mowery and Rosenberg 1982:165).

The definition of technological discontinuity in this study is similar to that of Foster (1986). This means that a technological change exists only if the technology moves from one S-curve to another.

5.4   Redefinition of the Dominant Design

Although Anderson (1988) studied float glass in his research, he did not regard Pilkington's float glass as a dominant design. He mentioned two reasons why float glass did not emerge as the dominant design: the patent consideration (Anderson 1988:149 and 351) and the fact that Pilkington did not settle on a standard (Anderson 1988:149). For these two reasons Anderson did not try determine whether float glass fulfilled the rest of the operational definition of dominant design, i.e. having more than a 50 % market share of the new flat glass production installations during the three consecutive years after introduction into the industry. I have to think three additional aspects to Anderson analysis before I can test whether float glass could have been regarded as a dominant design in the U.S. flat glass industry. These three additions are: (1) to keep the plate glass and sheet glass industries separate that is to define the industries with five-digit SIC codes, (2) to include patent considerations in the model and (3) to redefine a "narrow range of configuration." These additions are addressed next.

First, I have to keep the plate glass (SIC 32112) and the sheet glass (SIC 32111) subindustries separate and second I have to remember that float glass was developed within the plate glass industry as the following quotations require:

> Float glass would only be launched on the world if it could replace plate glass. If float glass merely provided an improved sheet glass, it would occupy a peculiar position between two glasses with well-established positions, one of which (sheet) had very low margins (Quinn 1977:8).

This (plate glass) was a more heavily capitalized and more fiercely competitive branch of the industry and was only entered into after much debate between the two branches of the family. In the early years of Cowley Hill, however, the profits of sheet were required to support plate, for soon after the new factory got into full production, foreign imports of plate glass began to rise even faster than those of sheet, driving prices down and down. Between the wars, however, it was the turn of Pilkington plate glass to come to the rescue of its sheet; and it was plate, not sheet, which led on float. In retrospect, therefore, the decision to move into plate glass in the 1870s can be seen as one of the greatest importance (Barker 1994:33–34).

Second, I must recall the patent considerations, which according to Anderson (1988:351) precluded the emergence of a dominant design. The success of float glass put a very powerful weapon into Pilkington's hands. Prestigious and profitable plate glass manufacturing lines were obsolescent by the early 1960s. Pilkington realized that lower fixed capital costs might encourage newcomers into the plate glass industry, in which very high capital costs had previously deterred newcomers. Pilkington's policy on float glass licensing was, therefore, to license the existing major producers and not to put them at a disadvantage by licensing newcomers (Barker 1994:87). One extreme case in Pilkington's licensing policy might have been the West German manufacturer, Detag, which was also in the 'plate' glass business with a different raw material. Detag used thick sheet glass as its raw material. According to its policy, Pilkington did not granted a license for float glass technology to Detag (Spoerer et al. 1987:178–179; Barker 1994:118). Here I have to again keep the industries separate (that is to use five-digit SIC codes) and see Pilkington's licensing policy guarding the plate glass industry,

By 1968 float glass had become competitive with sheet glass in certain thicknesses (Quinn 1980:55). However, float glass was much superior to sheet glass, and therefore sheet glass manufacturers began to deluge the company with license applications. Nevertheless, a long development process was required before the float glass process could make a full range of commercial thicknesses (Quinn 1989:835–836).

Third, the definition of dominant design needs some further discussion. The operational definition (Anderson 1988:107) of dominant design is a single configuration or narrow range of configurations that accounted for at least 50 % of the new process installations or new products in at least three consecutive years following a discontinuity. Anderson uses the definition of "narrow range of configuration" liberally. Sometimes it appears to be broad and other times narrow. For instance the Fourcault machine, one type of drawing machine, was regarded by Anderson (1988) to be exactly the same (the most important is the width and thus the capacity) during the 20 year period from 1917 to 1937 (1988:342), thus fulfilling the requirements of dominant design. However, according to the following quotation the width of the drawing machine was increased from 52 to 90″ (or 100″), which represents 73 (or 92) % increase in capacity.

The first Fourcault Drawing Machines used in the United States were designated as 148 cm (52-inch ribbon) and 183 cm (72-inch) wide glass sheet. The 148 cm machines were originally used for drawing sheet glass thinner than single strength (0.090 inch).

With increased production requirements and improved technology throughout the 1930s, these were gradually replaced with 90-inch ribbon machines and eventually with machines drawing 100-inch wide ribbons of glass. Today, flat-drawn sheet glass ranging from 0.040 to 0.248 of an inch in thickness is successfully produced on the machines (Mowrey 1967:249).

Anderson also stretched the term "narrow range of configuration" in his analysis of the cement industry. The computerized long kiln was introduced in 1960 and it achieved the status of a standard (or dominant design) in 1965 (Anderson 1988:160). However, in the analysis, the size or the length and thus the capacity of the kilns was allowed to vary during the 1960s from 500 to 600 ft, i.e. 20 %, the process control (the process control discontinuity) was allowed to change, and multiple suppliers were allowed (1988:237). Anderson writes (1988:49):

> As noted before, dominant designs emerge for both technical and institutional reasons. Standards arise because organizations and markets demand them. Producers will not forever endure the uncertainty that accompanies design competition; for the firm, creative destruction is healthy only when competitors' products are being destroyed. Customers, too, seek to reduce uncertainty; they want to purchase a reliable product of known capabilities. The need for a dominant design is especially critical when compatibility is an issue…

In a sense, customization of a float glass line meets the following compatibility requirements extremely well, i.e.:

1. to build a totally new float glass plant,
2. to build a float glass line alongside a plate glass line or
3. to convert the established plate glass installation to a float glass line,

mentioned in Anderson's last quotation. The float glass technology was/is such a strong standard (depending not only on the size or scale of the production line) that it allowed customization of the production line. In huge projects of this kind it was/is nearly impossible to require deliveries of the same line for 8–10 years. The process control technology (thanks to the semiconductor), for instance, made such huge strides during that period (1962–1972). The change in the thickness (thinner float glass was required) available for float glass also required modifications of the new lines. In other quotations Anderson (1988:149) says:

> In another invention (float glass), the inventor did not manufacture in the United States, and did not settle on a standard. Instead, the firm installed a custom line at each plant of each licensee, so that no two installations were quite the same (Anderson 1988:149)

This is a narrow understanding of 'standard.' One cannot compare individual float lines with each other as, for instance, video cassettes from the same standard (VHS or Beta) can be compared. Throughout the 1960s, computerized process control was one of the most rapidly advancing aspects in the industries (Quinn 1969) and thus, also for float glass manufacturing (Ames 1974; Skriba 1971). At its introductory stage the float glass process (Pilkington 1963; see Appendix 4) required computerized process control. This technological discontinuity could

have been called 'process control discontinuity' like that of the cement industry
(Anderson 1988:267). In my opinion, float glass was a much stronger standard than
the computerized long kiln in the cement industry, because float glass was controlled by a single company, as Anderson mentioned (1988:136)

> Pilkington completely controlled the diffusion of the float glass technology until 1973,
> when its largest licensee (PPG Industries) developed its own patented process (which it
> refused to license to competitors) (Anderson 1988:136).

Process control was also emphasized in the trade journals. TGI (Uusitalo
1995b:144) took the following quotation from the annual report of the Honeywell
company:

> Technology is significantly altering the nature of industrial instrumentation. Digital
> techniques and computers are becoming increasingly important in process control,
> requiring a high degree of capability in systems engineering, as well as a broad range of
> instrumentation and control equipment.

The following quotation of the inventor of float glass, Sir Alastair Pilkington
explains where the industry standard was going to be (in technology, not in the
scale or output).

> People easily fail to understand that the greatest secret about a new process is not how to
> do it, but that it can be done (Quinn 1977:11).

Float glass was implicitly regarded as an industry standard in the quotation
from Henry Ford II's dedication speech in April 1967 at the inauguration of Ford's
second float glass line:

> The technology you are looking at here was developed in England. It is called the Pil-
> kington process (Ford Motor Co. 1967:309).

As a result of this discussion float glass is regarded as being 'within narrow
range of configuration'. This discussion also provides support the definition of
'within narrow range of configuration' mentioned in Chap. 3, i.e. technology is
within narrow range if it is licensed by one company (in this case by Pilkington).

5.5 Float Glass: Was it a Dominant Design in the Flat
Glass Industry?

This chapter considers whether float glass can be regarded as the dominant design
in the flat glass industry. The test is carried out in both subindustries, the plate
glass and the sheet glass, in the two different markets, the United States and
Europe (see Fig. 1.6). The operational definition of a dominant design in the
Anderson study was as follows: a technological change will be a dominant design
at the moment when it has more than a 50 % market share in all new process
installations for three consecutive years. In order to examine whether float glass
became a dominant design, I must first identify the introduction of float in each

particular industry in each market, and second, calculate the new flat glass manufacturing process installations in both subindustries and in both markets. In the plate glass industry I calculate either installed plate glass lines or installed float glass lines, while in the sheet glass industry I calculated either installed sheet glass drawing machines/plants or installed float glass lines.

5.5.1 Did Float Glass Emerge as the Dominant Design in the United States?

The Plate Glass Industry (SIC 32112). Float glass was commercially introduced in the United States by Pittsburgh Plate Glass in spring 1964. PPG's first float glass line was built in the U.S. (Cumberland, Maryland, referred to as PPG1 in Fig. 5.3). The last plate glass line ever installed in the United States (and in the world) was American St. Gobain's (ASG) line at Greenland, Tennessee in 1962. Since the introduction (in 1964) of float glass in the United States all manufacturing process installations in the plate glass industry were float glass lines. In three consecutive years, 1964–1966, five new float lines were installed in the U.S. plate glass industry. Both PPG and LOF started production with two lines: PPG (in Cumberland/PPG1, the first in 1964 and Crystal City/PPG2 in 1966) and LOF (in Lathrop/LOF1 and Rossford/LOF2) and Ford with one line in Nashville (Ford1) in 1966 (see also Appendix 5). No other manufacturing process was even considered. Thus the market share of float glass process installations in plate glass line installations was 100 % during 1964–1966.

The Sheet Glass Industry (SIC 32111). In 1971 TGI (Uusitalo 1995b:146) wrote that Ford, Guardian and Pilkington Canada had started to sell float glass at the price of sheet glass. In May 1970 Ford shut down its last sheet glass plant. In October 1970 Guardian sold its first float glass. In December 1970 Pilkington started production on its second float glass line in Canada. According to the above information, one can conclude that float glass was introduced in the window glass industry in North America in 1970. Charles K. Edge from PPG Industries has reached the same conclusion:

> A process which, in 1965, was seen to be capable of displacing plate glass, was by 1970, clearly capable of supplanting sheet glass as well (Edge 1984:714/4).

PPG's factory in Fresno, California in 1967 and in Owen Sound, Ontario, Canada in 1968 were the last Pennvernon (drawn glass) window glass plants built in the United States and Canada, respectively. In 1971 PPG (Uusitalo 1995b:147) increased the sheet glass capacity (i.e. installed new Pennvernon drawing machines) of its Fresno plant. Guardian and Combustion-Engineering (C-E) were the first companies outside the plate glass industry to be granted a float glass license. In 1970–1972, i.e. in three consecutive years after introduction of float glass in the sheet glass industry, five lines started to produce float glass (Guardian in Carleton/Guardian1 in 1970, Ford in Nashville/Ford4 1971, Combustion-

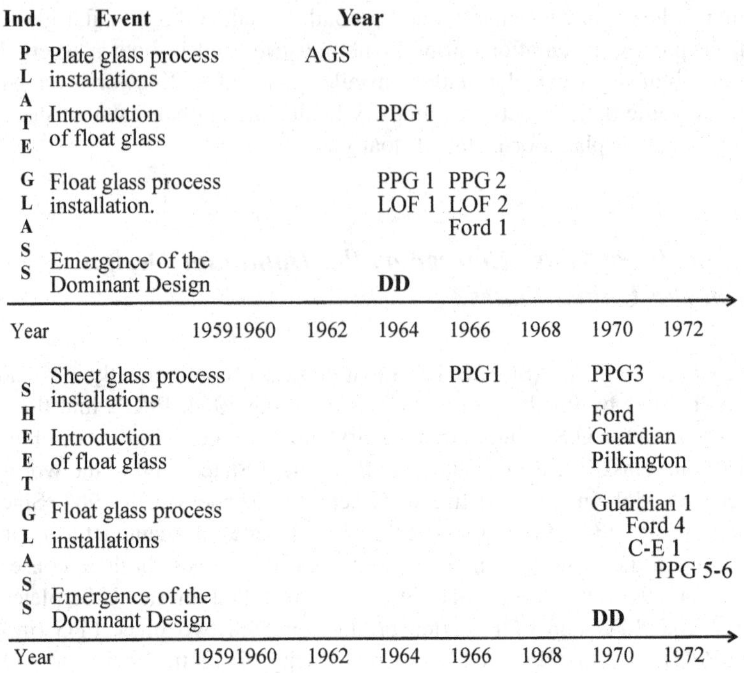

Ind.	Event	Year							
P L A T E	Plate glass process installations	AGS							
A T E	Introduction of float glass			PPG 1					
G L A S S	Float glass process installation.			PPG 1	PPG 2				
				LOF 1	LOF 2				
					Ford 1				
S S	Emergence of the Dominant Design				**DD**				
Year		1959 1960		1962	1964	1966	1968	1970	1972

Ind.	Event	Year							
S H E E T	Sheet glass process installations					PPG1		PPG3	
E E T	Introduction of float glass							Ford Guardian Pilkington	
G L A S S	Float glass process installations							Guardian 1 Ford 4 C-E 1	PPG 5-6
S S	Emergence of the Dominant Design							**DD**	
Year		1959 1960		1962	1964	1966	1968	1970	1972

Fig. 5.3 Events in the United States crucial for testing Anderson's model

Engineering (C-E) in Florette/C-E1 in 1971 and PPG two lines in Carlisle/PPG 5-6, see also Appendix 5). Thus the market share of float glass installations in the sheet glass industry was more than 50 % in 1970–1972.

All the events in these two branches described above are shown in Fig. 5.3. As a conclusion from this information I can say that float glass became the dominant design in both the plate glass and sheet glass subindustries (in 1964 and in 1970, respectively) in the U.S. market. In Fig. 5.4 the concept of dominant design (Anderson 1988) is applied in the illustration of the evolution of the flat glass industry in the 1960s and 1970s in the U.S. market. In Figs. 5.1 and 5.2 the technological discontinuities in both the U.S plate glass and the U.S. sheet glass industries are shown. In Fig. 5.1 there are four discontinuities, the continuous casting process (introduced in the United States in 1923), the twin grinding technology (in 1951), the twin polishing technology (in 1962) and the float glass (in 1964) processes, which were introduced in Sect. 4.1. In Fig. 5.2 the last two discontinuities in the sheet glass industry, the draw and float glass processes, were also discussed earlier. The first two, the Lubbers machine and the improved Lubbers, are mechanized glass blowing machines invented by American Window Glass company (Pilkington 1969a). In Figs. 5.1 and 5.2 the Y-axis does not have a scale interpreting how much the performance measure was increased after the discontinuity, i.e. the vertical dimension has symbolic meaning only. These figures

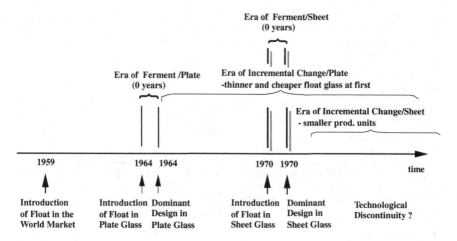

Fig. 5.4 Application of the modified dominant design concept to the diffusion of float glass into the flat glass industry in the United States

only illustrate the introduction time of technological discontinuities in the U.S. plate glass and sheet glass industries. These figures can be compared with that of Anderson (Anderson 1988; Figure C-1 in Appendix C).

5.5.2 Did Float Glass Emerge as the Dominant Design in Europe?

The Plate Glass Industry. Float was introduced in the plate glass subindustry in 1959 by the inventor, Pilkington (referred to in Fig. 5.5 as CH1). 3 years later, in 1962, Pilkington started the second line in Cowley Hill (CH3). During the period 1962–1964 four float glass lines, Pilkington in 1962 and in 1963 in Cowley Hill (CH3 and CH4), Glaverbel in 1965 in Moustier (Glaverbel), and St. Gobain in 1965 in Italy/Pisa (St. Gobain1), started production. As was mentioned earlier, St. Gobain's subsidiary in the U.S. installed the last plate glass line in 1962. During the 1960s not a single plate glass line was even considered in Europe.

The Sheet Glass Industry. In the 1960s at least, Detag, Delog and PPG from the U.S. invested in Europe in new sheet glass machines (referred to as Detag, Delog and Pennitalia in Fig. 5.5). By 1968 float glass had become competitive with sheet glass in certain thicknesses (Quinn 1980:55). At this time float glass was introduced in the European sheet glass market (referred as Pilkington). In West Germany the sheet glass and plate glass subindustries had been very close to each other before the introduction of float glass.

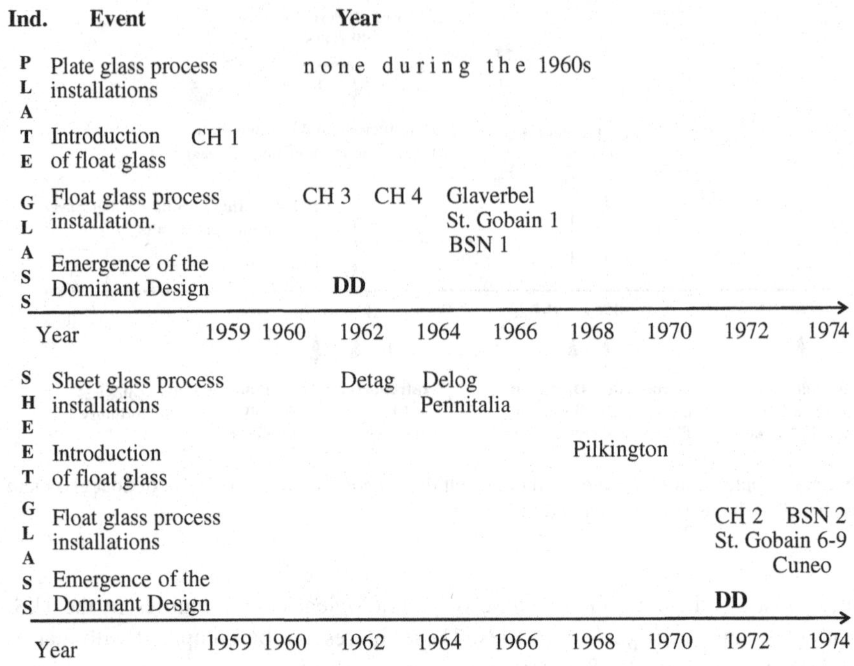

Ind.	Event	Year			
P **L** **A** **T** **E**	Plate glass process installations Introduction CH 1 of float glass	none during the 1960s			
G **L** **A** **S** **S**	Float glass process installation. Emergence of the Dominant Design	CH 3 CH 4 **DD**	Glaverbel St. Gobain 1 BSN 1		

Year	1959 1960	1962	1964	1966	1968	1970	1972	1974

| **S** **H** **E** **E** **T** | Sheet glass process installations

Introduction
of float glass | Detag | Delog
Pennitalia | | Pilkington | | |
| **G** **L** **A** **S** **S** | Float glass process installations

Emergence of the
Dominant Design | | | | | | CH 2   BSN 2
St. Gobain 6-9
Cuneo
DD |

Year	1959 1960	1962	1964	1966	1968	1970	1972	1974

Fig. 5.5 Events in Europe crucial for testing Anderson's model

There TDS (thick drawn sheet glass) had been polished so that it could be used as a raw material for more demanding applications such as mirrors and automobile safety glass. Good quality thick sheet glass was already a threat to plate glass in the 1930s. Float glass started to replace thick sheet glass in the late 1960s. In 1972–1974 four float glass lines (BSN in Gladbeck/W-G in 1974 and 1976 referred to as BSN2 and BSN3 in Fig. 5.5 and St. Gobain in Stolberg/W-G in 1974 referred to as St. Gobain Vereira di Verante in Cuneo/Italy in 1974 referred to as Cuneo) started production compared with zero drawing machines. BSN's subsidiary, Flachglas, the successor of the companies Detag and Delog, which had not belonged to the earlier plate glass subindustry, had two of the new plants. All the events described above are shown in Fig. 5.5. I can conclude from this information that float glass became the dominant design both in the plate and sheet glass subindustries (in 1962 and in 1972, respectively) in the European market (see Fig. 5.6).

In Chap. 3 important concepts, performance parameter, technological discontinuity and dominant design of Anderson's cyclical model of technological change model, were redefined in their model. The more strict definition with five-digit SIC codes of industry made it possible to better evaluate used technologies and international context of the subindustries an thus refined the crucial concepts in the model. Then the modified concept of dominant design was applied separately

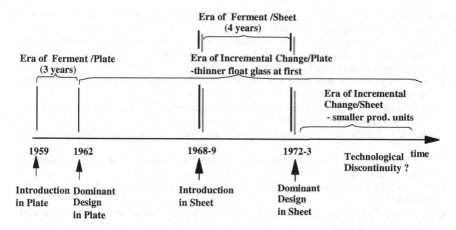

Fig. 5.6 Application of the modified dominant design concept to the diffusion of float glass into the flat glass industry in Europe

within both subindustries, plate glass (SIC 32112) and sheet glass (SIC 32111) of the flat glass industry (SIC 3211) in the United States and Europe. Float glass turned out to be the dominant design in both U.S. and European plate glass and sheet glass industries.

References

Ames BC (1974) Automation: key to greater productivity in glassmaking. Am Glass Rev February:11–12

Anderson P (1988) On the nature of technological progress and industrial dynamics. Unpublished PhD dissertation, Columbia University

Anderson P, Tushman LM (1990) Technological discontinuities and dominant designs: a cyclical model of technological change. Adm Sci Q 35:604–633

Anderson P, Tushman LM (1991) Managing through cycles of technological change. Res Technol Manage 34:26–31

Arrow K (1962) The economic implications of learning by doing. Rev Econ Stud 29:155–173

Barker TC (1994) An age of glass. Pilkington: the illustrated history. Boxtree, London

Daviet J-P (1989) Unemultinational a la Française. Histoire de Saint-Gobain 1665–1989. Fayard, Paris

Derclaye M (1982) La diffusion d'une innovation technique, le "float-glass" et ses consequences sur le marche du verre plat. Annales de Sciences Economiques Appliquees 38(1):133–148

Edge KC (1984) Section 11, flat glass manufacturing processes (update). In: Tooley F (ed) Handbook of glass manufacture, 3rd edn., vols I and II, Books for the glass industry division. Ashlee Publishing Co, New York, pp 714/1–21

Ford Motor Co. (1967) Newest Float Glass Plant. The Glass Industry, June, pp 309–313

Foster R (1986) Innovation: the attacker's advantage. Summit Books, New York

Katz D, Kahn R (1966) The social psychology of organizations. Wiley, New York

Kinkead G (1983) The End of ease at Pilkington's. Fortune, 21 March, pp 90–92, 94, 96

Miller D, Friesen PH (1982) The longitudinal analysis of organizations: a methodological perspective. Manage Sci 28(9):1013–1034

Mowery DC, Rosenberg N (1982) Technical change in the commercial aircraft industry, 1925–1975. In: Rosenberg N (ed) Inside the black box: technology and economics. Cambridge University Press, Cambridge, pp 163–177

Mowrey FW (1967) Technological progress in the flat glass industry. The Glass Industry, May, pp 247–249

Perry RC (1984) Float glass process: a new method or an extension of previous ones? The Glass Industry, February, pp 17–19, 31

Pilkington A (1963) The development of float glass. The Glass Industry, February, pp 80–81, 100–102

Pilkington A (1969b) Glass and windows chance memorial lecture. Chemistry and Industry, 8 February, pp 156–162

Quinn JB (1969) Technology transfer by multinational companies. Harvard Bus Rev 47(6):147–161

Quinn JB (1977) Pilkington Brothers LTD. The Amos Tuck School of Business Administration, Dartmouth College

Quinn JB (1980) Strategies for change. Logical incrementalism. Dow-Jones Irwin, Homewood

Quinn JB (1989) Pilkington Brothers P.L.C. In: Mintzberg H, Quinn JB (eds) The strategy process, concepts, contexts, cases. Prentice Hall, New Jersey

Rosenberg N (1982) Inside the black box: technology and economics. Cambridge University Press, Cambridge

Saint-Gobain (1965) Compagnie de Saint-Gobain, 1665–1965: Livre du tricentenaire, Paris

Skriba DA (1971) How a computer and glass communicate. The Glass Industry, December, pp 442–445

Spoerer M, Busi A, Krewinkel HW (1987) 500 Jahre Flachglas, 1487–1987 Von der Waldhütte zum Konzern, (in German, 500 year old flachglas, 1487–1987, from a small business to a concern). Karl Hofmann Verlag, Schorndorf

Tushman LM, Anderson P (1986) Technological discontinuities and organizational environments. Adm Sci Q 31:439–465

Tushman LM, Romanelli E (1985) Organizational evolution: a metamorphosis model of convergence and reorientation. Res Org Behav 7:171–222

Chapter 6
Conclusions

Abstract This chapter has four sections. First, the main findings of the research are summarized. The very broad definition of the industry with four-digit SIC (very broadly) seems to have two consequences: a) the omission of two pertinent aspects from the flat glass industry, entirely separate subindustries and the international nature of the subindustries, and b) the confusion of effectiveness and efficiency. Float glass was first effective in the plate glass industry requiring high quality (expensive) and thick flat glass. Not until float glass was thin and cheap enough it was capable of entering the sheet glass industry. In order to track these fundamental aspects three pertinent concepts, performance parameter, technological discontinuity and dominant design of the original model had to be redefined. Second, theoretical implications mainly discuss the challenges and the risks of entirely quantitative methods with very shallow data. Third, managerial implications deal the convergence of two industries. The dramatic change was hard to recognize since so many industry actors gave confusing signs. The U.S. based PPG was still 13 years after the introduction of float glass advocating sheet glass. Fourth, further research suggests studies of the further elaborations of the cyclical model of technological change such as the organizational determinants of technological change typology as well as the complexities involved in the innovation and diffusion processes of simple nonassembled products.

Keywords Effectiveness · Efficiency · Risk of quantitative method

Chapter 6 has four sections. The first section summarizes the main findings of the research. It seems that the definition of industry with four-digit SIC (very broadly) has caused the omission of important aspects from the flat glass industry such as two entirely separate subindustries, the international nature of the subindustry, the confusion of effectiveness and efficiency. Float glass was first effective in the plate glass industry requiring, high quality (expensive) and thick flat glass. Not until float glass was thin and cheap enough it was capable to enter the sheet glass industry. To track these fundamental aspects three pertinent concepts, performance parameter,

O. Uusitalo, *Float Glass Innovation in the Flat Glass Industry,*
SpringerBriefs in Applied Sciences and Technology,
DOI: 10.1007/978-3-319-06829-9_6, © The Author(s) 2014

technological discontinuity and dominant design of Anderson's (1988) model had to be redefined. The second section, theoretical implications mainly discuss the quantitative methods and its challenges in the regard of the depth of the information. Next section, managerial implications deals with matters relating the convergence of two industries. The change was hard to recognize since so many industry actors gave confusing signs. The U.S. based PPG 13 years after the introduction of float glass was still advocating sheet glass. Further research, the fourth, suggests the study of the further elaboration of the Anderson and Tushman's cyclical model such as the organizational determinants of technological change typology as well as the complexities involved in the innovation and diffusion processes of simple nonassembled products.

6.1 Summary and Main Findings

The main findings of this study were discussed in Chap. 5. Chapter 5 included the analysis and testing of cyclical model of technological change proposed by Anderson (1988) and Anderson and Tushman (1990, 1991), Barker (1994). It seems that the consequences of the definition of industry with four-digit Standard Industry Classification (SIC) code has caused different view to those made by the industry definition of five-digit SIC code.

The cornerstone of this study is the cyclical model of technological change proposed by Anderson (1988) and Anderson and Tushman (1990). It seems that the definition of the subindustries, the plate glass and the sheet glass industries, with five-digit SIC codes brought new insights in the analysis. The difference in the products (optical quality and the thickness, see Fig. 6.1) and the international context were taken in the account. In the 1950s sheet glass and plate glass were in different places on the optical quality and price continuum. Sheet glass was cheap and low quality while plate glass expensive and high quality. As was mentioned float glass was development to replace expensive plate glass. Pilkington, as the innovator, was really careful in introducing float glass. It did not like to have situation were float glass would have occupied a peculiar position between plate glass and float glass. A creative imitator (Drucker 1985) would have easily taken float glass to its right place for replacing plate glass.

In its introduction in 1959 float glass was much cheaper to produce than plate glass. However, the optical quality of float glass was not all the time good. It was not consistent. The available thickness (6.5 mm) of float glass was the most used thickness in the car industry. Thus, float glass was not yet effective in 1959 on the market. Anyway at that time it was too expensive for the sheet glass industry and it was not available in thicknesses (2 and 4 mm) needed there. After three years in 1962 the float manufacturing process produced good quality glass to replace plate glass. It was competitive (effective) on the market because its manufacturing process was efficient. Float glass was a right product (effective = do right things) for the market and its manufacturing process was

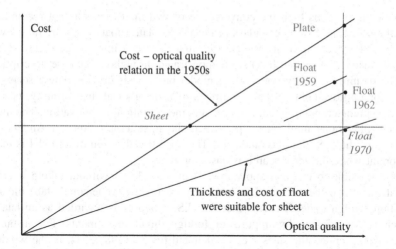

Fig. 6.1 The development of Float Glass in the 1960s on the base of manufacturing cost and thickness

economic (efficient = do things right). However, in the sheet glass industry it was effective not until thinner thicknesses were available (effective) and its costs (efficient) were low enough. The United States in 1970 float glass was both effective (right thickness) and efficient (low price) in the sheet glass industry which did not pay the price difference between high quality float glass and low quality sheet glass.

The main findings from the re-examination of the model are divided into two groups; redefinitions and consequences. Based on the more thorough qualitative empirical analysis of this study Anderson's model could be improved and modified. The modification includes redefinition of three fundamental concepts: (1) performance parameter, (2) technological discontinuity and (3) dominant design. The present research provides a performance parameter which measures the effectiveness of an organization. Technological discontinuity is defined simply as a jump from an existing technology to a novel one. This definition clearly leaves out technological changes caused by economies of scales. The redefinition of dominant design (and a narrow range of configuration) increases our understanding of industry standards in process technologies.

6.2 Theoretical Implications

The definition of an industry in a research is also of the utmost importance. In Anderson's case the strict definition of an industry by four digit SIC codes broke first, the tight link between flat glass and its two subindustries, plate glass and sheet and second, the link between plate (float) glass manufacturers and safety glass manufacturers. Thus both the convergence of two industries was not recognized and the importance of safety glass producers to plate (float) glass producers was ignored. However, the convergence (the first link) and importance of the safety glass industry (the second link) are demonstrated throughout the case description.

The definition of key concepts in model building is crucial. There seems to room for improvements in the definitions of technological discontinuity, performance parameter and dominant design, and the definition of innovation. The more exact explanations might have prevented the misinterpretations in Anderson's analysis of the U.S. flat glass industry. The operationalization of a model is more important when further research is based on it.

Is it possible to test and apply models of these kinds without taking international trade into the account? In his literature review Anderson (1988) did not mention foreign trade. It seems that the U.S. economy is regarded as an island, which is not affected by imports or foreign direct investments. For instance, Anderson compiled the sales, i.e. the demand figures (1988:351), in the window glass industry from the U.S. Census of Manufacturers, although 25 % of the demand was imported.

The importance of patent protection has been emphasized by many authors (Tushman et al. 1986; Tushman and Romanelli 1985; Pavitt 1987; Nelson and Winter 1982). Pavitt (1987) had even suggested to Anderson that the degrees of appropriability by innovators should be included in any models of competitive dynamics (Pavitt 1987:126). Moreover, Nelson and Winter (1982) say that the analysts of diffusion have not in general been cognizant of licensing policy in the diffusion process. However, patent protection is a difficult element in management studies. For instance, Anderson (1988) in his research treated patent protection as a nuisance, disrupting the patterns he expected to observe and generating unwanted variance in the results (Anderson 1988:201–202). The patent treatment has been especially problematic in the case of the Pilkington float glass process. For instance, Ray (1974), who has studied the diffusion of this innovation, had the following to say:

> Pilkington protected their innovation by a world license. Nobody could start a float line, quite apart from the necessary know-how, without an agreement with the license-holder, whose licensing policy could thus influence adoption of the process elsewhere. As far as one can assess it, this licensing policy was liberal. Negotiations were often lengthy, but were conducted on a 'first come, first served' basis. Understandably, some queuing was inevitable for the negotiations, but this did not cause any major delay and the dates of the license agreements reflect approximately the order of the licensees' decisions to apply for a license. (Ray 1974:210).

As was mentioned Pilkington's licensing policy was not liberal. On the contrary, Pilkington did not grant licenses outside the plate glass industry. The strict treatment of patent consideration is one of the major theoretical contributions in the present study.

When the two subindustries (the plate glass and the window glass branches) are handled separately, as they should be, some conclusions of Anderson's hypothesis testing can be challenged. Float glass was eventually the dominant design in both subindustries as mentioned in Chap. 5 (see also Figs. 5.1, 5.2).

The proper operationalization of a model is crucial to model building. The present research suggests many new ideas concerning definition of the key concepts. The present study also suggested two consequences of a radical technological change. A gradual convergence of two industries and the consolidation of an industry after a radical competence-destroying technological discontinuity were suggested.

The present study suggested that if a researcher in a longitudinal, historical study relies almost entirely on statistical data (for instance census figures), certain nuances of the empirical phenomenon are likely to be missed or misunderstood. As was mentioned, Pavitt (1987) was somewhat suspicious of Anderson's data. Rosenberg (1982) and Miller and Friesen (1982) are along the same line. The author of the present study agrees with Pavitt's findings, since in Anderson's work, the quality of the data seems to be inadequate due to the sole use of quantitative material (the U.S. Manufacturing Census statistics).

One of the hypotheses in Anderson's study discusses the length of the era of ferment. The hypothesis is as follows:

> The era of ferment following a competence-destroying discontinuity is longer than the era
> of ferment following a competence-enhancing discontinuity. Anderson (1988:161).

This hypothesis was supported in the study (Anderson 1988:142). The test had six of both types of competence cases. The average length of the era of ferment was 11.16 years for competence-destroying discontinuities and 8.0 years for competence-enhancing discontinuities. However, two of the most recent dominant designs (float glass in the plate glass and the window glass branches) are missing from the competence-destroying side. In both cases the era of ferment was extremely short (zero for both in the U.S. market). If these new results for float glass are incorporated into the test, the new average length of the era of ferment is 8.37 for competence- destroying discontinuities. When the average lengths of both discontinuities, i.e. 8.0 for competence-enhancing and 8.37 for competence-destroying are compared, the support for this hypothesis is missing.

6.3 Managerial Implications

The marketing policy of the sheet glass producers seemed to be rigid and it had multi- level distribution. These levels were difficult to bypass. In fact, this was found to be one reason for the increasing imports of sheet glass into the United States in the late 1950s and throughout the 1960s. The local U.S. flat glass producers accused the importers of dumping. Many European sheet glass manufacturers exported throughout the 1960s to the United States. Their agent even participated in the tariff hearings. The European flat glass manufacturers even preferred exports over domestic sales.

In the early 1970s, when float glass entered the U.S. sheet glass industry, exports from Europe dropped dramatically. Overcapacity also began in Europe and European manufacturers were in the same situation as the U.S. sheet glass manufacturers a few years earlier. Imports of float glass (also sheet glass) rose and finally European local sheet glass manufactures lost their domestic market. It seems that the European sheet glass manufacturers could not foresee the U.S. situation (i.e. the increasing sheet glass imports).

According to Clark (1983), if competition is the driving force behind industrial evolution then uncertainty about the preferences and technology is its fundamental prerequisite (Clark 1983:108). Clark (1983) continues, saying that without uncertainty on both sides of the market, evolution would not occur. On the producer's side of the market there must be a nontrivial collection of potential product technologies with different internal capabilities and uncertainty about which set of technologies will satisfy market requirements.

It seems that Anderson's (1988) model probably lacks the competing technologies and thus also the uncertainty in the sheet glass industry. In my opinion this lack is present because his model seems to ignore the nature of the gradual convergence of the subindustries. His model shows that float glass was introduced in the sheet glass industry as early as 1963 and thus, there was neither competition between technologies nor uncertainty in the market place. However, float glass was in reality introduced in this subindustry much later than in the plate glass subindustry. On the contrary, uncertainty among small manufacturers was extremely high, especially because two of the largest flat glass manufacturers, PPG from the U.S. and Asahi from Japan, were still developing the existing drawing processes.

PPG is an exciting example. The firm quickly adopted float glass in its plate glass business. This was mainly due to the superior and cost efficient technology of float process compared with that of plate glass, but partly due also to the foreign competition (ASG's plate line inaugurated in 1962). However, PPG's story in its sheet glass business was different. In a way, PPG supported its self-esteem as late as 1973 by continuing development of its own Pennvernon process. Only after having developed its own floatPPG glass process did the company rapidly convert sheet glass production to the floatPPG process. Naturally, the small sheet glass manufacturers knew nothing about the internal debate on technologies (the existing Pittsburgh and the new float) at PPG.

When the production costs of sheet glass and float glass were compared at the end of the 1960s and the early 1970s, the licensing costs (i.e. the amortization of the lump sum and the royalty) were taken into account (Friedrich, 1975:68–69). When the licensing costs were included, the production costs of a float glass plant (capacity of 500 tons/d) were 10 % higher than those of a sheet glass plant (with three drawing machines) (Friedrich 1975:68–69). However, Pilkington as the owner of float glass technology had no licensing costs which meant that its own production costs were equal or lower than those of a sheet glass plant. The production costs which included licensing costs thus left implicitly Pilkington's export possibilities out of consideration. It is important to notice that this kind of situation can take place in any other industry as well.

The high productivity and good process control potential of the float glass process were emphasized in the evaluations of different technologies (Berg 1984:56). However, I believe that few persons in the sheet glass business realized the ultimate productivity of a float glass line. Pilkington's third line, CH2, in Cowley Hill produced one million tons (650 tons/day) of float glass in less than four years and three months with only two production stops, for 72 and 48 h. (TGI April 1975:34).

Pilkington's licensing policy was really working well (see Wierzynski 1968) and it is an excellent example how to treat difficult markets (see more Uusitalo and Grønhaug 2012).

Float glass innovation provides an interesting example of the effects of technological discontinuity on industry structure. The flat glass industry is a reasonably compact and distinct industry, which also facilitates research, and this case is an accurate and detailed description of the changes over a long period. Nowadays, however, technological changes are more rapid. It seems that this case serves as a means to forecast the consequences (in 1960–1980 for the float glass innovation) of the coming technological changes in other industries if the time scale is shrunk from 20 years to approximately five or even less years.

6.4 Recommendation for Further Research

Tushman and his associates have used the cyclical model of technological change (Anderson and Tushman 1990; Barker 1994) as a basis for their studies (McGrath et al. 1992; Tushman and Rosenkopf 1992; Rosenkopf and Tushman 1994; Tushman and Murmann 1998). The author of the present study has noticed little published material critically evaluating Anderson's and Tushman's (1990), Barker (1994) model.

The further elaboration of Anderson and Tushman's dominant design model, the organizational determinants of technological change typology (Tushman and Rosenkopf 1992; Rosenkopf and Tushman 1994) is not examined in this study and second, of the complexities also involved in the innovation and diffusion processes of a simple nonassembled product, flat glass.

Competing technologies in the same organization are an interesting phenomenon. This case illustration provided two clear examples of this. PPG, a U.S.-based manufacturer, and Asahi, a Japanese manufacturer, both invested in R&D for both technologies. On the one hand, PPG and Asahi licensed float glass technology respectively in 1962 and in 1964, but, on the other hand, they devoted huge amount of resources to the old drawing processes in the late 1960s and the early 1970s. However, after having invented the own float glass technology which did not violate the patents possessed by Pilkington, PPG quickly replaced the existing sheet glass plants in the late 1970s. The case of competing technologies in the same firm, probably in PPG, would be an interesting research topic.

References

Anderson P (1988) On the nature of technological progress and industrial dynamics. Unpublished PhD Dissertation, Columbia University

Anderson P, Tushman LM (1990) Technological discontinuities and dominant designs: a cyclical model of technological change. Adm Sci Q 35:604–633

Anderson P, Tushman LM (1991) Managing through cycles of technological change. Res Technol Manage 34:2631

Barker TC (1994) An age of glass. Pilkington the illustrated history. Boxtree, London

Berg B (1984) Det Stora Glaskriget (The Glass War). Glasmästeribranschens sevice AB

Clark K (1983) Competition, technical diversity, and radical innovation in the U.S. Auto industry. In: Rosenbloom RS (ed) Research on innovation, managementand policy. vol. 1. JAI Press, Greenwich, pp 103–149

Drucker P (1985) Innovation and entrepreneurship. Pan Books, London

Friedrich H (1975) Der technische Fortschritt in der Glaserzeugung. Eine Untersuchung über die Auswirkung des technischen Fortchritt auf den Strukturwandel in der Flachglasindustrie (Bochumer Wirtschaftswissenschaftliche Studien Nr. 7), Bochum

McGrath RG, MacMillan IC, Tushman LM (1992) The role of executive team actions in shaping dominant designs: towards the strategic shaping of technological progress. Strateg Manag J 13:137–161

Miller D, Friesen PH (1982) The longitudinal analysis of organizations: a methodological perspective. Manage Sci 28(9):1013–1034

Nelson RR, Winter SG (1982) An evolutionary theory of economic change. The Belknap Press of Harvard University Press, Cambridge

Pavitt K (1984) Sectoral patterns of technical change: towards a taxonomy and a theory. Res Policy 13(6):343–373

Pavitt K (1987) Commentary on Chapter 3. In: Pettigrew A (ed) The management of strategic change. Basil Blackwell, Oxford, pp 123–127

Ray, GF (1974) Float glass In: Nabseth L, Ray GF (eds) The diffusion of new industrial processes, an international study. Cambridge University Press, p. 319

Rosenberg N (1982) Inside the black box: technology and economics. Cambridge University Press, Cambridge

Rosenkopf L, Tushman LM (1994) The coevolution of technology and organization. In: Baum J, Singh J (eds) Evolutionary dynamics of organization. Oxford University Press, Oxford, pp 403–424

Tushman ML, Murmann JP (1998) Dominant designs, technology cycles and organizational outcomes. Res Organ Behav 20:231–266

Tushman LM, Romanelli E (1985) Organizational evolution: a metamorphosis model of convergence and reorientation. Res Organ Behav 7:171–222

Tushman LM, Rosenkopf L (1992) Organizational determinants of technological change: toward a sociology of technological evolution. Research in organizational behavior. JAI Press, Greenwich, pp 311–347

Tushman LM, Newman W, Romanelli E (1986) Convergence and upheaval: managing the unsteady pace of organization evolution. Calif Manag Rev 29(1):29–44

Uusitalo O, Grønhaug K (2012) Service-dominant logic and licensing in international b2b markets. J Bus Mark Manage 4:265–284

Wierzynski GH (1968) The eccentric lords of float. Fortune 90–92:121–124

Appendix 1
Operationalization of the Main Concepts

TGI = The Glass Industy (trade journal)

Concept	Process to identify	The Most pertinent data sources
Technological discontinuities in the flat glass industry	Comprehensive study of the industry and company histories	Bain (1964) Barker (1977a, 1994),
		Berg (1983), Skeddle (1977)
		Daviet (1989), Friedrich (1975),
		Hamon (1988), Spoerer et al. (1987)
Classification of the subindustries	Comprehensive review of the empirical material	Bain (1974) Barker (1977a, 1994),
		PPG (1983), Borup (1.2.1993), Simpson (1961, 1963)
The state of the plate glass and sheet glass subindustries in the 1950s	Extensive study of the flat glass industry histories	Bain (1964), Frederiksen (1994)
		Pincus (1983), Barker (1977a, 1994),
		Tooley (1984), Skeddle (1977)
Flat glass manufacturing technologies and technology barriers	Review of the literature on flat glass manufacturing technology	Bain (1964), Persson (1969), Doyle (1979),
		Pilkington (1963), Takahashi et al. (1980),
		Borup (int. 1.2.93), Pincus (1983),
		Vincent (1960, 1962), Skeddle (1977)

(continued)

O. Uusitalo, *Float Glass Innovation in the Flat Glass Industry*,
SpringerBriefs in Applied Sciences and Technology,
DOI: 10.1007/978-3-319-06829-9, © The Author(s) 2014

(continued)

Concept	Process to identify	The Most pertinent data sources
The float glass process and sciences / technologies involved	Review of the literature on float glass	Pilkington (1963, 1969 a, b, 1971, 1976), Earle (1967), Edge (1984), Havard (1976)
The development of the float glass process	Review of the literature on the float glass R&D work	Pilkington (1963, 1969 a, b, 1971, 1976), T. C. Barker (int.),Grundy (1990),
Identification of the performance measure	Identification of the key product attributes optical quality, thickness and price	see previous item, Artama (interview), Derclaye (1982), Quinn (1977), Skeddle (1980)
Determination of the branch in which float glass was invented	Investigation of Pilkington's invention process for float glass	Skeddle (1977), Barker (1994), Quinn (1977), Wierzynski (1968), TGI (a review of it), Bain (1964), Borup (interview)
Identification of the commercial introduction time (i.e. innovation) in both industries and markets	Review of company histories, TGI, statistics and interviews	Barker (1977a, 1994), Quinn (1980) Grundy (1990), TGI (Feb. 1971:82), Friedrich (1975), Spoerer et al. (1987)
Identification of new process - installations (i.e. market shares for determining whether the process emerges as a dominant design)	Comprehensive review of the empirical material	TGI (a thorough review) Grundy (1990), Friedrich (1975) Berg (1983)
Definition of a narrow range of configurations within float glass	To understand the core of an industry standard, the need for customization of a float glass lines, the role of process control and licensing	Pilkington (1963), Quinn (1977), Skeddle (1977), TGI (June 67:309) Ashton (1969), Skriba (1971) Review of TGI

(continued)

(continued)

Concept	Process to identify	The Most pertinent data sources
Identification of Pilkington's licensing policy	Comprehensive review of the empirical material plus the case writing, interviews	Wierzynski (1968), Barker (1994) Quinn (1977), Lowry (1982) Spoerer et al. (1987), Kinkead (1982)
Identification of regionalization and/or globalization of the industry	Calculation of the number of flat glass producers and FDIs (either in the flat, safety glass industries or distribution outlets)	Barker (1977a, 1994), Daviet (1989) Hamon (1988), Spoerer (1987) Berg (1983), Frederiksen (1974)
Identification of concentration in the flat glass industry	See the previous item	
Identification of other external forces (such as value added products, foreign trade, FDIs, customs duties, etc.)	Comprehensive review of the empirical research material with a special emphasis on economic events	Artama (interview), news clippings TGI (a thorough review) Bain (1964), Skeddle (1977)

Appendix 2
Distribution for and Comments
to the Licentiate Thesis, B-156, HSE

Pilkington Plc		
Sir Antony Pilkington	Chairman	1980–1995
	MD, vice MD, etc.	1967–1980
	(Flat Glass Division)	
Ian Burgoyne	Curator (Museum)	
PPG Industries Inc.		
Frank A. Archinago	Vice President, Glass	
London School of Economics and Political Science		
T. C. Barker	Professor	
Manchester Business School		
Alan W. Pearson	Professor	
	Working with Pilkington	1951–1961
Lahden Lasitehdas		
Jonas Borup (also earlier	MD	1989–1993
interviews)	Deputy MD	1988–1989
	Administrative Dir.	1971–1988
	Department Manager	1967–1971
Kurt Lindqvist	MD	1982–1988
	Production Director	1975–1982
	R&D Director	1972–1975
	Production Engineer	1959–1972
Bo Sandberg	MD	1976–1982
Arne Möller	Chairman of the board	1976–1991
Lamino		
Erkki Artama	Managing Director	1975-
	Technical Director	1965–1975
Riihimäen Lasi		
Antti O. Kolehmainen	Managing Director	1965–1976
	Deputy MD	1959–1965
	Accountant Manager	
Ministry of Trade and Industry		
Bror Wahlroos	Secretary General	1969–1992

MD = Managing Director

O. Uusitalo, *Float Glass Innovation in the Flat Glass Industry*,
SpringerBriefs in Applied Sciences and Technology,
DOI: 10.1007/978-3-319-06829-9, © The Author(s) 2014

Appendix 3
Growth of Float Glass in Advanced Countries 1960–1976

Plants constructed and demolished
According to published record the "float revolution" has resulted in the following change in numbers of clear flat glass tanks operating in the United States and Europe:

	Sheet		Plate		Float	
	1960	1976	1960	1976	1960	1976
North America	37	12	15	1[a]	-	29
U. K.	7	1	2	Nil	1	4
Continental Europe	53	17	17	Nil	-	17

[a] Has not operated for 4–5 years and is expected to be demolished in 1977

In summary this means that the construction/demolition program over 16 years since float glass was announced to the world has been:

	Furnaces		
	Constructed Float	Demolished Sheet	Demolished Plate
North America	29	25	14
U. K.	3	6	2
Continental Europe	17	36	17
TOTAL	49	67	33

Source Modified from Exhibit IV in Quinn 1977:30

O. Uusitalo, *Float Glass Innovation in the Flat Glass Industry*,
SpringerBriefs in Applied Sciences and Technology,
DOI: 10.1007/978-3-319-06829-9, © The Author(s) 2014

Appendix 4
The State of the Float Glass Process in 1963

1. There is no limitation on the output from a float bath. The speed at which the ribbon is formed is only determined by the melting capacity of the furnace.
2. The variable costs for converting molten glass into a finished ribbon are remarkably low and are only a small fraction of the equivalent part of the plate process.
3. The width of the ribbon can be easily altered at any time within the limits of the float bath and the lehr. This can be important in reducing warehouse loss by matching the ribbon to the orders which are being cut.
4. It produces a finished ribbon at the end of the lehr which encourages the use of an automatic warehouse.
5. The process leads itself to long, trouble-free runs. Pilkington has already run for 14 months without stopping and knows that a single run can be designed for the whole life of a furnace.
6. As the volume of molten glass remains constant at all thicknesses, the speed becomes very high on thin glass and we have had to design lehrs for much higher speeds than we have ever used before.

Source Based on Pilkington (1963)

O. Uusitalo, *Float Glass Innovation in the Flat Glass Industry*,
SpringerBriefs in Applied Sciences and Technology,
DOI: 10.1007/978-3-319-06829-9, © The Author(s) 2014

Appendix 5
Plants with a Float Glass Process License

The company	Plant location	In operation	Capacity	Inv.
(licensing year)		(under glass)	t/day or Msq.ft/Y	$ mill
United States				
PPG Ind.	1. Cumberland, MD.	Mar. 1964	300/50	
(1962)	2. Crystal City, Mo.	Jan. 1966	300/50	
	3. Meadville, Pa.	Oct. 1968	300/50	
	4. Meadville, Pa.	May 1970	300/50	20
	5-6. Carlisle, Pa twin	Apr. 1972		50
Float$_{PPG}$	7-8. Wichita Falls, Tex.	Nov. 1974		
"	9. Fresno, Calif.	1977		20
"	10. Mount Zion, Ill.	1981	650	
LOF	1. Lathrop, Calif.	Sep. 1964		
(1963)	2. Rossford, O.	Sep. 1966		18
	3. East Toledo, O.	Jun. 1969		20
	4. Ottawa, Ill. 3/69	Apr. 1970		20
	5. Rossford, O. 5/69	Aug. 1970		21
	6. Laurinburg, NC	Jun. 1973		40
	7. Laurinburg, NC	1980		
Ford Motor Co.	1. Nashville, Tenn.	Mar. 1966		
(1964)	2. Dearborn, Mich.	Apr. 1967	300/50	30
	3. Nashville, Tenn.	Jun. 1968		
	4. Nashville, Tenn.	Mar. 1971		
	5. Tulsa, Okla	1974		
	6. Tulsa, Okla	1976		
Guardian	1. Carleton, Mich.	Oct. 1970	350/58	10
(1971)	2. Carleton, Mich.	Oct. 1973		
	3. Kingsburg, Calif.	1977	500	30
	4. Corsicana, Texas	1980	500	
Combustion-	1. Florette, Pa.	Dec. 1971		10
Engineering	2. Cinnaminson, NJ	1974		25
(1970)				

(continued)

O. Uusitalo, *Float Glass Innovation in the Flat Glass Industry*,
SpringerBriefs in Applied Sciences and Technology,
DOI: 10.1007/978-3-319-06829-9, © The Author(s) 2014

(continued)

The company	Plant location	In operation	Capacity	Inv.
ASG/AFG Ind.	1. Greenland, Tenn.	July 1973	450	
(1971)	2. Greenland, Tenn.	1981		
	3. Greenland, Tenn	1984	600	35
Fourco Glass Co.	1. Jerry Rund., W.Va	Nov. 1974		
(1973)				
Canada				
Pilkington	1. Scarborough	Jan. 1967	357	30
Ontario	2. Scarborough	Dec. 1970	500	32
PPG Float$_{PPG}$	1. Owen Sound	1978	550	30
(license)		(under glass)	t/day	£ mill.
Europe				
Pilkington	1. CH1, UK	Jan. 1959		
	2. CH3, "	Jun. 1962		
	3. CH4, "	Mar. 1963		
	4. CH2, "	Sep. 1972	650	12
	5. Halmstad, Sweden	1976	550	67
	6. UK5, UK	1981		122
	7. Lahti, Finland	1987		
BSN	1. Boussois, France	Feb. 1966		
(1962)	2. Gladbeck, W-G	Mar. 1974		
	3. Gladbeck, W-G	1976		
	4. Weiherhammer	1979	700	
Glaverbel (1962)	1. Moustier, Belgium	Mar. 1965		
St. Gobain	1. Pisa, Italy	Dec. 1965		
(1963)	2. Porz, W-G	Feb. 1966		
	3. Aviles, Spain	Feb. 1967		
	4. Auvelais, Belgium	Feb. 1970		
	5. Herzogenrath, W-G	1971		
	6. Chantereine, France	1972		
	7. Catologne, Spain	1973		
	8. Caserta, Italy	1973		
	9. Stolberg, W-G	1974		
	10. Flovetro, Italy	1979		
Veireria di	1. Cuneo, Italy	1974		
Vernante (1972)				
Guardian	1. Luxguard, Luxemb.	1981		
PPG	1. Salerno, Italy	1983		
Asahi	1. De Maas, Holland	1983		

CH = Cawley Hill

Appendix 6
A Comparison of the U.S. and European Markets

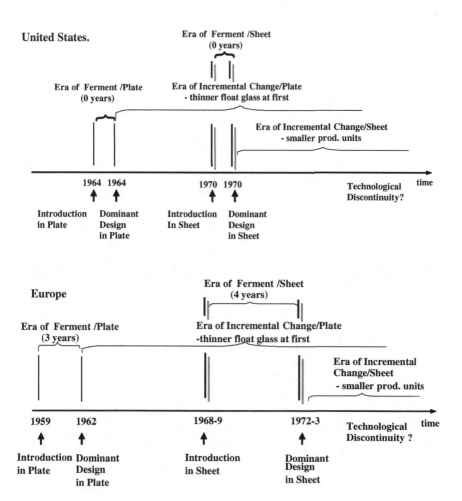

O. Uusitalo, *Float Glass Innovation in the Flat Glass Industry*,
SpringerBriefs in Applied Sciences and Technology,
DOI: 10.1007/978-3-319-06829-9, © The Author(s) 2014